Instructor's Resource Manual for

ENVIRONMENTAL GEOLOGY

An Earth System Science Approach

Dorothy Merritts

W. H. FREEMAN AND COMPANY
NEW YORK

SUPPLEMENTS EDITOR: *Patrick Shriner*
ASSOCIATE EDITOR: *Robert Christie*
PROJECT EDITOR: *Erica Seifert*
ADMINISTRATIVE ASSISTANT: *Ceserina Pugliese*
COMPOSITION AND DESIGN: *Christopher Wieczerzak*

ISBN 0-7167-2762-5

Printed in the United States of America

First Printing, 1998

Contents

Part 1
Fundamental Concepts: How Humans View and Study Earth

Part 2
Solid Earth Systems

Part 3
Fluid Earth Systems

Part 4
Energy and Change in Earth Systems

Preface

Some educators might wonder why we have used an Earth system science approach in a textbook for an environmental geology course. Although we adopted this approach long ago in our own teaching of the course, our reasons for doing so were articulated recently in a public document produced from a workshop on Earth science education. In the fall of 1996, a panel of 48 Earth and space scientist educators, including senior author Dorothy Merritts, met at the American Geophysical Union's headquarters in Washington, D.C., under support from the National Science Foundation and the Keck Geology Consortium, to discuss the future of undergraduate Earth science education. The panel concluded that Earth system science provides such a strong, unifying context in which "to demonstrate the interrelationships between all components of the Earth system and humanity" that they felt compelled to recommend that colleges and universities adopt such an approach at all levels of the curriculum in the Earth and space sciences.

In 1997, the panel released a document entitled "Shaping the Future of Undergraduate Earth Science Education: Innovation and Change Using an Earth System Approach" (available on the Web at http://earth.agu.org/sci_soc/spheres/). In this document, they justify their recommendation:

> The Earth and space sciences are in the midst of a major revolution. New insights into the dynamics of the Earth system and the relationships between its many subsystems make this an exciting time of new discovery. . . . We are coming to understand that all of humanity impacts, and is impacted by, the Earth system. This emerging awareness of the interrelationships between natural and social systems makes it imperative that all people have a fundamental understanding of the Earth system. (p. 7)

> This exciting broadening and integration of the Earth and space sciences is a revolution in thinking like that produced in plate tectonics 30 years ago. . . . Our new understanding requires that we move away from traditional, compartmentalized thinking. (p. 13)

We agree wholeheartedly with the panel's recommendations and find it exciting to teach students how the atmosphere, hydrosphere, biosphere, solid Earth, and pedosphere are connected, as well as the ways in which they interact. Because environmental issues usually involve several, or sometimes even all, of these systems, an Earth systems approach is necessary to understand these issues fully.

Instructors can help students to assimilate and understand the textbook material by explaining to them how its content is organized. The *Environmental Geology* textbook is divided into four parts. Part I, Fundamental Concepts: How Humans View and Study Earth, includes Chapters 1 through 3. These chapters provide the background on scientific methodology, Earth systems, and geologic time that are necessary to consider the remaining chapters. Chapters 4 through 10 focus on Earth systems, all of which can be divided into two basic types: those that are solid (Part II, Solid Earth Systems) and those that are fluid (Part III, Fluid Earth Systems). The fourth and final part of the book, Part IV, Energy and Change in Earth Systems, explains the driving forces that cause change, evidence of environmental change in the geologic record, the history of change on Earth, and efforts to predict future conditions on Earth.

SAMPLE SYLLABUS FOR A CLASS THAT MEETS TWICE A WEEK

Week 1   Environmental Geology, Earth System Science,   Chapter 1
         and the Scientific Method

         Critical Issues in Environmental Geosciences
         (Population, Resources, Wastes, and Hazards)

Week 2   Origin of Planet Earth   Chapter 2

         Earth Systems, Cycles, Feedback, and Global Change

Week 3   Geologic Time: Relative and Absolute   Chapter 3

         Radioactivity: Radiometric Dating and Radon Gas

Week 4   Elements, Minerals, and Rocks   Chapter 4

         Plate Tectonics and the Rock Cycle

Week 5   Ore Deposits, Depletion of Mineral Resources,   Chapter 5
         and Environmental Impact of Mining

         Volcanoes and Volcanic Eruptions

Week 6   Earthquakes: Causes, Hazards, and Risks

         Weathering: Producing Sediments and Soils   Chapter 6
         from Rocks

Week 7   The Soil Resource: Soil Erosion and Conservation

         Mass Movement: Processes, Hazards, and Control

Week 8   The Hydrologic Cycle and Surface Water Systems   Chapter 7

         Floods: Recurrence Intervals, Hazards, and Control

Week 9   Stream/Watershed Restoration and Wetlands

         Water Quality: Regulations and Protection

Week 10   Groundwater: Occurrence, Storage, and Flow   Chapter 8

         Groundwater Hazards and Pollution

SAMPLE SYLLABUS FOR A CLASS THAT MEETS THREE TIMES A WEEK

OUTLINE FOR EACH CHAPTER

With the exception of Chapter 12, which combines Chapter 12 and 13 from the textbook, each chapter of this resource manual matches the same chapter in the textbook and contains five categories of material, described below.

Chapter Objectives

The chapter objectives are drawn from the opening of each chapter in the textbook, but our choice of these objectives is explained here in several paragraphs provided for the instructor's use.

Chapter Outline

Here we provide an outline of each chapter to be readily available to the instructor during course preparation.

Suggested Lecture Outline

Next, we present an outline containing headings and subheadings that could provide a framework for class presentations. These headings are designed to pull together two or more parts of a chapter, for we realize that no instructor can cover all the topics in any given chapter during a few lectures, let alone one or two. The headings are also designed to help the instructor identify key ideas and questions that are addressed in the textbook.

Suggested Lecture and Discussion Topics

We begin each of our classes or topics with a hook—examples that will get students interested in a particular topic or help them to see the topic's relevance to their lives. We give one or two examples of such hooks for each chapter. Some chapters include more detailed case study examples in this section as well.

Demonstrations and Case Studies

We conclude with examples of demonstrations and case studies appropriate for each chapter. All the demonstrations are simple and require little equipment or preparation time. All can be done during class by the instructor, sometimes with help from students. The case studies are similar to those in the textbook but are included here to give instructors interesting examples to provide in class that will be new to the students. Note that additional case studies are sometimes included in the suggested lecture and discussion topics as well.

CHAPTER 1

Introduction to Environmental Geosciences

CHAPTER OBJECTIVES

One of our most important responsibilities as educators is to help Americans understand and appreciate science as a way of knowing about our world. Some 15 million people are enrolled in our colleges and universities, and only about 15 percent of those students are majors in science or engineering. Most of the rest, however, are required to take one or more science classes to get a degree. As stated by Bruce Alberts, current president of the National Academy of Sciences, this 85 percent of our students

> will become the citizens who determine—by their understanding and appreciation for the nature and values of science—both the vitality of our nation and the future of our scientific enterprise. It would be fine if all Americans knew about plate tectonics, or the way that cells divide. But it is much more important that they understand what science is (and what it is not) and how its central values—honesty, generosity, and respect for the ideas of others—have made possible the rationalization of human experience that underlies all human progress.
> (*Science Teaching Reconsidered*, 1997, p. v.)

Chapter 1 calls upon our special responsibility to further scientific literacy. It aims to address the following questions:

What distinguishes scientific thinking from other types of thought?

Why is the scientific method the most effective strategy yet devised for learning about physical events?

What are the limitations of scientific thinking?

How is science used in the interest of conserving Earth's resources and avoiding environmental risks?

What is the advantage of Earth systems science for understanding the global environment?

CHAPTER OUTLINE

1.   Science as a Way of Knowing
      A.   The Scientific Method
      B.   Approaches to Scientific Reasoning
      C.   The Link between Science and Technology
      D.   The Limitations of Science

2.   History of Science
      A.   Galileo, Newton, and the Changing World View
      B.   Darwin, Wegener, and Changing Views of Time and Place

3.   Evolution and Environment
      A.   The Evolution of Humans
      B.   Human Effects on Environment

4.   Critical Issues in Environmental Geosciences
      A.   Human Population Growth
      B.   Resources and Sustainable Development
      C.   Pollution, Wastes, and Environmental Impact
      D.   Natural Disasters, Hazards, and Risks

5.   Environmental Geosciences and Earth System Science
      A.   Linked Environmental Systems
      B.   Earth Systems Science

SUGGESTED LECTURE OUTLINE

The headings listed below are related to topics addressed in Chapter 1 and provide an alternative structure by which to present the material during classes and/or discussion periods. Items marked with one asterisk are treated in the following section on suggested lecture and discussion topics; those marked with two asterisks are treated in the section on demonstrations and case studies.

Science as a Way of Knowing
      What is science?*
      Why is science important to our lives?
            Examples of recent scientific issues in the news
      What are scientists like?*
            Media images of scientists*
      What is the scientific method?*
      What is not science?

History of Science
> How have advances in science affected our interactions with the environment?

Evolution and Environment
> When did our species appear on Earth, and what has influenced its evolution since then?
> How has human evolution affected our ability to change Earth's environments?

Critical Issues in Environmental Geosciences
> Population growth
>> Is our population growing exponentially?
>> What is causing the rapid rise in population?
>> How many people can the Earth support?
> Resources: scarcity, depletion, and sustainability
>> Renewable, nonrenewable, and perpetual resources**
> Pollution, wastes, and environmental impact
>> What kinds of pollution and waste do we produce?
>> What is the magnitude of human impact on the environment?**
> Natural disasters, hazards, and risks
>> What are the most significant geologic hazards?
>> How does risk differ from disaster?

Earth Systems Science—A New Worldview That Bridges Many Fields of Study
> What is Earth systems science?*
> Why is Earth systems science a useful way to approach environmental issues?*

SUGGESTED LECTURE AND DISCUSSION TOPICS

Science as a Way of Knowing

One way to begin an inquiry-based, interactive approach on the first day of class is to ask students to compose a concise definition of science. Give them three minutes to formulate the definition in one or two sentences in their notebooks. Ask a few students to read their definitions aloud, then lead a short discussion about them. For example, if one student says, "Science is a large collection of facts about natural phenomena," you can ask the rest of the class if they agree with that statement (most might agree!). You might counter that scientists often explore things about which they know very little (for example, the interior of the Earth earlier in the 20th century; outer space), so science must be more than a collection of facts. You might also assert that the students use scientific reasoning and methodology in their daily lives, even though they are not scientists.

After a short discussion of various students' definitions, write the following statement on the board (or display it as an overhead) and ask the students if it provides an adequate definition of science.

> "A means by which humans try to understand the world, their surroundings, and their existence."

This definition gets at the notion that science is more than just a body of knowledge. Science is also a *way of knowing*. Of course, the definition could also apply to other human endeavors, such as religion or astrology. One could argue, for example, that humans attempt to understand their existence through religion and that the statement could be part of a definition of this way of human understanding.

Humans have developed various ways of interpreting their surroundings. In particular, they have searched for answers to what philosophers call the ultimate questions:

> Who am I?
> How did we get here?
> What is the meaning of life?
> What are good and evil?
> What happens to us when we die?

Science cannot answer all these questions, and it is worth discussing with the students that science—despite its power and acknowledged successes—does have limitations. For example, ask the students which of these questions science is unlikely ever to be able to answer. Most will agree that science will be able to answer only the second question.

At this point, another statement can be added to the definition:

> "An activity of creativity and imagination that requires discipline and self-criticism."

Certainly, this statement can apply to many human endeavors—including art as well as science. Students generally are not accustomed to viewing the practice of science as requiring creativity and imagination. We use this part of the discussion to share with students our own passion for using scientific reasoning, and we often show a few slides that demonstrate some of the types of work in which we engage. Sometimes we use examples of us doing field or lab work with students. Or we take students on a quick tour of our laboratory facilities. Our goal is to give students a glimpse of the life of a scientist. If most students take only one or two science courses, the professors of those courses might be the only scientists with whom they come in contact while in college. Those professors have a special opportunity to bring science to life for their nonscience majors.

To conclude the discussion of what science is, the definition given in the textbook can be restated:

> "The scientific approach to understanding our environment is based on the assumption that natural events have physical, and therefore ultimately knowable, causes. Scientific knowledge is acquired through the systematic and disciplined testing of ideas by careful experimental design, methodical data collection, and objective reasoning and analysis."

Another interesting discussion for the first or second class can be started by asking the students to describe three scientists from books or film. Typical responses from our classes include the Professor on "Gilligan's Island"; Dr. Frankenstein; and Emmet Brown, the eccentric inventor in the movie *Back to the Future* and its sequels. Ask the students how these characters portray scientists. Most students will recognize that scientists are often portrayed as nerds, fiends, or absent-minded geniuses. At this point, we like to share with our students examples of well-known scientists who have risked their lives for the sake of an important discovery. One example is Stan Williams, who was seriously injured while studying a volcano in Columbia. The story of Dr. Williams is described in the opening to Chapter 1 in the textbook.

Environmental Geosciences and Earth System Science

Another topic worth addressing at the beginning of the semester is the use of Earth system science as a way of investigating environmental processes. An excellent example that illustrates how various Earth systems are closely interconnected is found in Chapter 4 (Box 4.2, How Plate Tectonics Affects Species) and can be introduced here briefly.

Because of plate-tectonic motions, South America has been drifting northward from the South Pole for millions of years, and about 3 million years ago its northern tip converged with North America. This docking of the two continents enabled species from each landmass to come into contact for the first time in nearly 100 million years. North America and South America are fragments of Laurasia and Gondwanaland, respectively, and they have had different types of animals and plants since those landmasses split apart from Pangaea some 70 million years ago. Over the course of 70 million years, marsupial mammals continued to evolve on both continents, but placental mammals evolved only in Laurasia after it separated from Gondwanaland. As a result, placental mammals—which had some evolutionary advantages over marsupials—became dominant in North America, Europe, and Asia, and marsupials became extinct.

When North and South America collided, however, these two groups of mammals were once again able to interact. Many of the marsupials in South America became extinct; some migrated northward. Placental mammals from North America migrated southward, and some of them became extinct as well. By the 18th century, only one fragment of Gondwanaland—Australia—remained isolated from the northern landmasses and placental mammals. The introduction of rabbits by a farmer in the 19th century changed that isolation forever, and some 20 million rabbits now are the bane of Australian landowners.

While discussing this example with students, we show a computer animation of drifting continents (from Planet Earth, produced by Victor Schmidt). If you do not have this or similar software, overheads that illustrate the position of continents over the past few hundred million years can be projected. The goal is merely to make clear to students that species extinction is not a new phenomenon and in fact has occurred in the relatively recent geologic past as a result of moving landmasses. At this point, it is useful to refer to Figure 1.6 in the textbook, which illustrates that the number of plant and animal species generally has increased over the past 600 million years, even though there have been a number of episodes of large global extinctions. At present, the number of species in the world is decreasing as a result of human activities, and this decrease can be discussed in the context of biological diversity and species extinction revealed in the geologic record.

DEMONSTRATIONS AND CASE STUDIES

The Environmental and Social Consequences of Poor Mining Practices and Mineral Depletion

Only time will tell exactly how the world's increasing demand for finite resources will affect the environment, but it is certain that some areas will be completely depleted of their ore deposits, some ecosystems will be damaged and degraded by mining activities, and some groups of people will suffer immensely as a consequence. A recent example that illustrates all three of these results is phosphorus mining on Nauru, a small Pacific island (population: ~10,000) that is the world's smallest republic.

Phosphate deposits are very important to the fertilizer industry and are essential to sustaining high agricultural productivity. Since the early 1900s, about 1 percent of the world's phosphorus has come from several small Pacific islands, including Nauru. Like many islands in the South Pacific, Nauru is composed of the fossilized remains of ancient coral reefs. Over the past one to two million years, glaciers repeatedly retreated and advanced in the northern hemisphere, and global sea level rose and fell in response. Coral reefs grew upward during periods of high sea level, then died off and stabilized at much lower levels during periods of low sea level. Sea birds dropped their phosphate-rich guano (excrement) onto the emerging coral islands. Thus, Nauru's pinnacled surface of partly dissolved limestone was covered with a rich biological ore of phosphate, before most of it was mined in the twentieth century.

Nauru was annexed by Germany in 1888. In 1907, the Germans began to mine the phosphate. Australia seized the island during World War I and mined more than a third of the ore before Nauru achieved independence in 1968. The liberated Nauruans continued mining, but by 1995, four-fifths of the guano was gone.

The inhabitants of Nauru are now contemplating a future on some other island, because their own has become uninhabitable. An island once covered with tropical forest now is bleak, desolate, and dry, with only a few coconut trees along its coastline. The local climate has been altered because the exposed gray limestone pinnacles, up to 23 m high, have a much higher albedo (ability to reflect light) than the once-forested landscape. They also shed more rainwater as runoff into the ocean. The result is a sun-baked desert.

The inhabitants abandoned farming decades ago in favor of collecting the royalties from mining. (Nauruans have one of the highest per capita incomes in the undeveloped world.) They also altered their diet from local fresh fish and vegetables to imported canned meats and processed foods. The consequences have been dire. Despite their substantial monetary wealth, the Nauruans have compromised their health. Women live only to age 55 on average and men to age 50, due in large part to rates of obesity and diabetes that are among the world's highest.

Nauru exemplifies a hard lesson: Earth's recoverable mineral resources are finite, and careless mining can result in severe environmental and human consequences. The Nauruans are not wholly to blame for their plight. As an Australian colony from 1914 to 1947, and then a trust territory of Australia, Britain, and New Zealand until independence in 1968, Nauru can hold another nation responsible for beginning the ruin of its environment. In fact, Nauru successfully sued Australia in the International Court of Justice in 1989 for colonial-era ravaging. It received $75 million, which it hopes to use to cover the island with new soil and to replant trees and crops. If the islanders are unable to regenerate the island's ecosystem and develop a means of future sustenance for themselves, they will use their money to buy another island on which to live.

CHAPTER
2

Dynamic Earth Systems

CHAPTER OBJECTIVES

In our own teaching of environmental geology at the introductory level, we find that Earth system science provides for us much of what the theory of plate tectonics must have provided for those teaching physical geology from the 1960s through the 1990s: a clear, coherent framework in which to describe various geologic and environmental processes, events, and features and their connections to one another. (See the preface for our summary of why an Earth system science approach is useful.) We consider it important to teach students some fundamental concepts about a systems approach before treating the Earth itself as a system. For this reason, Chapter 2 is designed to teach both the general concepts of a systems approach and the ways in which Earth processes can be analyzed from a systems perspective.

With these goals in mind, Chapter 2 aims to

Examine the concept of systems and show why it is a powerful tool for understanding how Earth works.

Briefly review Earth's planetary evolution and the major systems that make up the planet.

Identify the forces that drive Earth processes and examine how feedback mechanisms either amplify or regulate them.

Look at ways in which the rock cycle and the hydrologic cycle circulate matter and energy through the whole Earth system over time.

CHAPTER OUTLINE

1. The Concept of Systems
    A.   Types of Systems
    B.   Dynamic Systems That Tend toward a Steady State

2. The Planetary Evolution of Earth
    A.   Origin of the Universe and Our Solar System
    B.   Differentiation of Earth

3. Earth's Environmental Systems
    A.   The Lithosphere
    B.   The Pedosphere
    C.   The Hydrosphere
    D.   The Atmosphere
    E.   The Biosphere

4. Earth's Energy System
    A.   States of Energy
    B.   Sources of Energy
    C.   Energy Budget of Earth
    D.   Human Consumption of Energy

5. Feedback Links among Earth Systems
    A.   Positive and Negative Feedback
    B.   Evidence of Global Feedback

6. Change, Cycles, and Earth System Dynamics
    A.   The Rock Cycle
    B.   The Hydrologic Cycle

SUGGESTED LECTURE OUTLINE

The headings listed below are related to topics addressed in Chapter 2 and provide an alternative structure by which to present the material during classes and/or discussion periods. Items marked with one asterisk are treated in the following section on suggested lecture and discussion topics; those marked with two asterisks are treated in the section on demonstrations and case studies.

Earth as a System
        What is a system?
        In what ways does Earth operate like a system?
        What are the major environmental systems on Earth?

Evolution of the Earth System
     Origin of the universe
     Origin of the solar system and its planets
     Origin of planet Earth and its early history (planetary differentiation)
     Differences between Earth and other planets**

Energy and its Importance to System Processes
     Earth's energy sources
     Earth's energy budget
     The importance of energy to life on Earth—the rise of human civilization and its reliance on energy

Cycles, Change, and Feedback in Earth Systems
     What is a cycle?
     How do matter and energy cycle through Earth systems?
     Examples of feedback and change in Earth's environmental systems*

SUGGESTED LECTURE AND DISCUSSION TOPICS

Examples of Feedback and Change in Earth's Environmental Systems

Emphasizing the links among Earth systems rather than just the specific systems provides students with a framework for understanding environmental processes. The textbook gives many examples of feedback and change in Earth's environmental systems. Useful places to look for these examples are identified in the following index entries: Cycles; Earth systems, and environmental change; Earth systems, feedback links; and Feedback mechanisms. Chapter 2 provides an introduction to change and feedback, as well as details of several specific examples of each.

We find it useful to point out to the students that feedback and change in Earth systems are ubiquitous, yet it is possible to evaluate them in terms of four criteria:

     The type of change that occurred
     The time scale over which it operated
     The nature of the event or events that caused the change
     The type and nature of the feedback mechanism associated with the change

In our lecture on Earth systems and change, we discuss three examples (two of which are described in the textbook) in terms of these four criteria. As we elaborate on each example, we construct the table shown in Table I2.1. Below, we provide the page numbers for the two examples in the textbook, plus supplementary notes for the third example. By presenting information in the form of the table and its four criteria, the instructor is helping students to evaluate material from their textbook as well as to relate it to new material. A fourth example that could be added to the table is the species swapping between North and South America that occurred when the two continents collided several million years ago and that resulted in extinction of many marsupials (see Box 4.2 in the textbook; see also the Suggested Lecture and Discussion Topics section of Chapter 1 of this manual).

To start the discussion of real examples of environmental change and feedback in Earth systems, we begin with a statement followed by a definition of Earth system science.

Statement: An amazing aspect of the Earth system is that a natural event occurring at one location can result in complex and far-reaching effects throughout the whole Earth. Earth system science provides a global perspective on phenomena that might not seem related but actually are.

Definition (slightly modified from page 25 of the textbook): Earth system science focuses on the interconnections among various Earth systems and the changes that occur in them over time. Earth system science views the whole Earth as a single, integrated system in which matter and energy are cycled through numerous subsystems, including the lithosphere, pedosphere, hydrosphere, biosphere, and atmosphere. Humans are a part of the whole Earth system, depending upon it for resources, affecting its environments, and responding to its changes.

Example 1 is explained in more detail on pages 51–52 in the textbook. Figure 2.18 in the textbook shows the positions of continents about 30 million years ago, when the Drake Passage opened between the southern tip of South America and the Antarctic continent.

Example 2 is not described in the textbook, so we will give a few additional comments here. Some scientists have suggested that formation of the Isthmus of Panama had something to do with onset of the world's current Ice Age because of the coincidence in timing between the two. Although it is difficult to test this idea, it is certain that the Gulf Current became established at this time and that ocean currents were able to flow unimpeded along the equator before the collision of the two continents. Once this flow was blocked, warm equatorial water was able to flow northward along the eastern coast of North America, as well as southward along the eastern coast of South America. In the southern hemisphere, however, warm water was blocked by the Antarctic circumpolar current. It might be useful to refer students to the map of world ocean surface currents in Figure 10.18 of the textbook.

Example 3 (the impact of Mount Pinatubo's eruption on global climate) is explained in more detail on pages 28–31 in the textbook. Figure 2.1 shows the distribution of volcanic ash and acid particles in the atmosphere immediately after the eruption and again two months later.

TABLE I2.1 Criteria for Assessing Environmental Change

CHANGE	TIME SCALE	POSSIBLE CASUAL EVENT(S)	FEEDBACK MECHANISMS
Example 1			
Cooling and formation of ice sheet on Antarctica	Millions to tens of millions of years (began ~40 m.y. ago)	Antarctic and South American Continents separate; a circumpolar current becomes established and deflects warm equatorial currents. In addition, mountain building in Antarctica results in additional cooling (7°C cooling per km rise in elevation).	Once ice formation begins, positive feedback mechanisms result. Ice has a large albedo relative to land, so it reflects more solar radiation, thus causing further cooling in Antarctica.
Example 2			
Cooling and formation of ice sheets in northern hemisphere	Millions of years (began ~4 m.y. ago)	South America migrated northward and collided with North America, forming the Isthmus of Panama and blocking eguatorial currents; the Gulf Current developed and carried warm water northward, resulting in increased evaporation (and, as a result, increased humidity and snowfall), perhaps leading to ice formation.	Same as example above for Antarctica.
Example 3			
Global cooling of about 0.5°C	Months to years	Mount Pinatubo erupted in June 1991 and released 30 billion kilograms of sulfur dioxide that came from relatively small amounts of sulfur dissolved in the huge volume of magma. Sulfur dioxide combined with water vapor in the atmosphere to form sulfuric acid droplets (aerosols). In combination with 4 cubic kilometers of ash, the droplets circled the globe and remained in the atmosphere for two and a half years. This debris increased Earth's albedo, thus reflecting solar radiation and causing cooling.	As volcanic ejecta filled the atmosphere and increased its ability to reflect solar radiation, the resultant cooling enabled more water vapor to form in the atmosphere, causing further formation of aerosols and thus a positive feedback.

DEMONSTRATIONS AND CASE STUDIES

The Geology of Other Planets

This book focuses on the geology of Earth, but the other eight planets in our solar system have interesting geological features and histories of their own. The study of other planets is fascinating in its own right, and it can also give us insights into the geology of our own planet. Here we'll take a look at Earth's two closest neighbors, Venus and Mars. Even though these planets are more Earthlike than any of the others, their current geology and their geologic history are quite different from Earth's.

Venus

Traditionally, Venus and Earth have been considered sister planets because of their similar diameter, mass, density, and composition. Venus, however, is covered by a thick, reflective cloud layer (Figure I2.1) that produces very high surface temperatures (about 460°C) and pressures (about 90 atmospheres). In addition, Venus has no satellites, nor does it have a magnetic field.

The atmosphere of Venus is composed mainly of carbon dioxide (about 98 percent). Below Venus's clouds, which are made up of sulfuric acid droplets and solid sulfur particles, the atmosphere contains about 0.1 to 0.4 percent water vapor and 60 parts per million of free oxygen. The presence of water and free oxygen indicates that Venus may have had abundant water early in its history, which was later lost.

Venus has a very pronounced greenhouse effect. Sunlight filtering through the clouds heats the surface, which radiates the heat back into the lower atmosphere. Rather than escaping to space, the heat is easily absorbed by the carbon dioxide in the atmosphere and radiated back to the surface. This process is called the greenhouse effect because it is similar to the warming of a greenhouse. The glass of the greenhouse lets in visible light but lets little heat escape. Earth has a much less extreme greenhouse effect. If it had none, its surface temperature would be well below freezing and the oceans would be a solid mass of ice. There is concern, however, that Earth's greenhouse effect may be increasing because of increased carbon dioxide released into the atmosphere by the burning of fossil fuels. The increased carbon dioxide could result in global warming, which would have serious effects on climate and weather patterns and would result in a small but significant rise in global sea level. Studying Venus's more extreme greenhouse effect can give us insights into what might happen here on Earth.

Venus and Earth share many geological surface features. There are three basic types of surface on Venus: lowland plains (20 percent), rolling uplands (70 percent), and highlands (10 percent). All three areas show evidence of extensive volcanism. Much of the volcanic activity on Venus occurs as basaltic eruptions that flood large areas. In addition, Venus has several large shield volcanoes (Figure I2.2), some of which are currently active.

Venus also displays unique forms of volcanism. One such form is pancake domes (Figure I2.3), which are almost perfectly circular, flat domes with steep sides. The domes are typically about 25 km wide and 2 km high and appear to be made of highly viscous lava erupted suddenly from a single vent. There are also rivers of extremely fluid lava; about 40 of them are longer than 100 km, and one is 7000 km long. In contrast, the longest lava channels on Earth extend only a few tens of kilometers.

The most distinctive volcanic features on Venus are coronae (Figure I2.4), large circular structures with a slightly raised interior surrounded by a low circular ridge and a trough. They are typically 200 to 500 km in diameter, although the largest (Artemis) has a diameter of 2000 km. Coronae are thought to be hot spots formed over mantle plumes that became inactive before they could form a true shield volcano. It appears that Venus is still volcanically active, but only in a few hot spots; for the most part, it has been geologically quiet for the past few hundred million years.

Impact craters are distributed randomly but uniformly over the surface. From the number of craters and their sharp, uneroded condition, we deduce that Venus was resurfaced about 300 to 500 million years ago. The majority of Venusian craters appear pristine because they were formed after this resurfacing, and there has been little geologic activity and weathering since to degrade and destroy the craters. There are far fewer small craters on Venus than on other planets, because small meteoroids vaporize or break up in the Venusian atmosphere before they reach the surface.

The highland regions have mountain ranges, volcanoes, and rift systems. Mountain ranges, long faults, and deep troughs indicate that horizontal surface movement has occurred. Among the most complex types of terrain identified on Venus are raised regions of the surface characterized by intersecting sets of faults and ridges. These regions, called tesserae from the Latin word for "tile," may result from long episodes of compression and extension of the surface.

Although its surface has been very geologically active, Venus appears to lack plate tectonics. Heat generated in the interior drives the abundant volcanic activity. Features found at plate boundaries on Earth also occur on Venus, such as deep asymmetrical troughs associated with subduction zones and rifts typical of spreading centers. But these features do not appear to link up in an integrated system of plates as on Earth.

Mars

The major constituents of the Martian atmosphere are carbon dioxide (95.3 percent), nitrogen (2.7 percent), and argon (1.6 percent). The average atmospheric surface pressure is less than 1 percent of Earth's, and it varies with season and elevation. The surface temperature fluctuates greatly between day and night and between seasons, but even at the equator temperatures are below freezing most of the time. The average temperature is about −53°C. Although thin and frigid, the Martian atmosphere is very active and complex. Mars and Earth have similar global atmospheric circulation patterns. In the Martian atmosphere, as in Earth's, warm air rises at the equator, moves poleward, is deflected to the east, and then descends at middle latitudes and returns to the equator. At middle to high latitudes, jet streams blowing from the west produce storm systems near the surface. Mars also has seasonal climate changes caused by solar heating and by the exchange of carbon dioxide between polar ice and frost and the atmosphere. The polar caps grow and shrink seasonally. The strong southern summer winds lift vast amounts of dust into great storms that cover the entire planet (Figure I2.5).

Although the overall geology of Mars is unique in the solar system, it shares some characteristics with Earth. Many geologic features on Mars are much larger than the same kinds of features on Earth, however. For example, the shield volcano Olympus Mons (Figure I2.6), the largest volcano in the solar system, is three times as tall (24 km) as Mount Everest (the tallest mountain on Earth) and 20 times as large in cubic content as Mauna Loa (Earth's largest shield volcano). Several factors contribute to the size of geological features on Mars. The lack of plate tectonics means that centers of volcanic activity persist longer in the same place, rather than moving with the changing plate boundaries as on Earth. These more static conditions allow volcanoes to grow for longer periods of time. The surface gravity of Mars is only about a third that of Earth, which allows volcanoes to grow much higher before they reach their limiting heights. Mars also has a much less erosive atmosphere and climate, so volcanoes are worn down much more slowly.

Olympus Mons is one of four huge volcanoes in the Tharsis region, an immense area uplifted by tectonic forces and marked by volcanism. Many geologists believe that the Tharsis bulge resulted from a large mantle plume. Tharsis straddles and overlies the two major terrains on Mars: the southern hemisphere consists of ancient cratered uplands; the northern hemisphere of younger, relatively uncratered, volcanic lowlands. How the surface of Mars came to be divided into these two disparate terrains is not completely understood, but it is thought that the crust in the northern hemisphere probably collapsed as a result of internal forces present billions of years ago.

Stretching nearly 5000 km from the Tharsis slopes out across the old uplands is the Valles Marineris system (Figure I2.7). The system is up to 8 km deep and 200 km wide in places. The deep linear canyons of Valles Marineris were produced by the same tremendous tectonic forces that produced Tharsis. Although these canyons were not carved by running water, geologists believe that they were widened by catastrophic outbursts of water, ice, and debris released from artesian aquifers.

The surface of Mars shows features that resemble dry riverbeds and gullies. These might have been created by rainfall and runoff, but they might also have been made by subsurface water that seeped to the surface. In either case, the channels are evidence that temperatures were warmer in the past—warm enough for water to exist in liquid form. The higher temperatures probably were caused by heat released from the interior and by the thicker atmosphere.

A small number of meteorites found on Earth are known to have originated on Mars. In August 1996, a NASA research team announced that they had identified organic compounds in a Martian meteorite found in the Antarctic. They also found several mineral features characteristic of biological activity and identified possible microscopic fossils of primitive, bacteria-like organisms. This evidence suggests that primitive life may have existed on Mars more than 3.6 billion years ago.

The announcement set off a wave of research on this and other Martian meteorites. Some researchers have supported the original conclusion, while others have remained skeptical. Exciting as this discovery is, the evidence so far does not establish the fact of extraterrestrial life beyond a doubt. Much work remains to be done before we can be confident of this extraordinary claim. Ultimately, we will probably need to send a probe to Mars to scoop up samples of the rock and soil there and return them to Earth for study.

CHAPTER 3

Geologic Time and Earth History

CHAPTER OBJECTIVES

A grasp of the immensity of geologic time is as fundamental to environmental geology as it is to physical geology. The fact that environments on Earth are changing continuously is central to understanding how to deal with many modern environmental issues. For example, the international conference on global warming held in Kyoto, Japan, in December 1997 was based on the premise that Earth's average global temperature is increasing because of human activities. Yet some scientists argue that Earth's climate has changed frequently in the geologic past, long before the human species even appeared on Earth, and it is difficult to predict what might happen in the future.

Indeed, at the scale of millions of years, Earth's average surface temperature has been declining for about 65 million years, since the warm climatic conditions of the Mesozoic era. Yet at the scale of millennia, Earth has been warmer than at present. Within the current interglacial episode of the past 18,000 years, average global temperature might have been greater than today at least once, during the mid-Holocene (about 6000 years ago). During the last major interglacial episode about 124,000 years ago, average global temperature probably was higher than today for 1000 years or more.

The purpose of this chapter is to provide students with an understanding of the immensity of geologic time, the frequency of environmental changes in Earth's past, and the ways in which scientists have been able to date past events and unravel Earth's history. The two final chapters of the book, Chapters 12 and

13, revisit the concepts of time and change in much greater detail and draw upon the material presented in previous chapters. In contrast, Chapter 3 provides the foundation necessary to go on to Chapters 4 through 13. Specifically, the chapter's objectives are to

Describe how geologists developed a relative time scale from fossil evidence and rock formations.

Distinguish between relative and absolute geologic time scales.

Examine the principles of radiometric dating that led to an absolute geologic time scale.

CHAPTER OUTLINE

1.   Time, Space, and Earth Processes
    A.   Time's Arrow
    B.   Using Earth's History to Predict the Future

2.   Relative Geologic Time
    A.   Leonardo's Interpretation of Fossils
    B.   Steno's Laws of Rock Layering
    C.   Fossils and the Geologic Column
    D.   The Geologic Time Scale
    E.   Evolution

3.   Absolute Geologic time
    A.   Radioactivity: Nature's Atomic Clock
    B.   Radioactive Decay of Unstable Isotopes
    C.   Radiometric Dating
    D.   The Immensity of Geologic Time

SUGGESTED LECTURE OUTLINE

The headings listed below are related to topics addressed in Chapter 3 and provide an alternative structure by which to present the material during classes and/or discussion periods. Items marked with one asterisk are treated in the following section on suggested lecture and discussion topics; those marked with two asterisks are treated in the section on demonstrations and case studies.

The Importance of Scale (Temporal and Spatial)
    The immensity of geologic time*
    Examples of the temporal and spatial scales of Earth processes
    Early scientific efforts to determine temporal scales**

The Science of Geology and Geologic Time
    How geologists view time: billions and billions of years
    The principle of uniformitarianism
    Absolute and relative time

Extracting Relative Temporal Information from Geologic Materials
    Stratigraphic sections and the clues they contain (see Box 3.2 in the textbook)*

Dating Geologic Materials with Radioactive Clocks
    Unstable isotopes and radioactive decay
        Example of radon gas
    Decay rates and half-lives
    Useful isotopes for radiometric dating

Layered Sediments and Relative and Absolute (Radiometric) Age Dating
    Temporal clues from Mount St. Helens (see Box 3.1 in the textbook)
    Temporal clues from Yucca Mountain (see Box 3.4 in the textbook)
    Reconstructing the dates of past earthquakes (see Slides 3.1 and 3.2 in the Slide Set)*

SUGGESTED LECTURE AND DISCUSSION TOPICS

The Immensity of Geologic Time

Many instructors like to use analogies to emphasize the immensity of geologic time. Figure 3.10 in the textbook presents the Earth's pattern of changing climates and life forms on a time scale laid out between a person's fully extended arms. This illustration is a useful starting point for discussions about geologic time. We have found that when students read the text that accompanies the figure, many of them actually extend their arms as they consider the time at which major events in Earth's history occurred.

One particularly effective analogy for a 50-minute class period is described by biologist Rob Reinsvold (University of Northern Colorado) in Gordon Uno's *Handbook on Teaching Undergraduate Science Courses: A Survival Training Manual* (University of Oklahoma Printing Services, 1997). Reinsvold tells his students that the classroom can be viewed as a time machine in which the students will begin class at the time of Earth's origin and end class at the present. At the time of major events in Earth's history, one of Reinsvold's teaching assistants blows a whistle or rings a bell, and Reinsvold then pauses to explain the event. In Table I3.1, we have summarized the time of major events in the evolution of Earth and its life forms. The approximate age of the event is given in the second column. The transformed time for a 50-minute period is given in minutes and hundredths of a second in the third column. The fourth column is the transformed time in minutes and seconds (for example, 4, 21 is 4 minutes, 21 seconds).

The impact of the demonstration becomes clear as class progresses. For the first 40 minutes, major events are few and far between. The first land plants don't appear until the last 5 minutes of class. Dinosaurs appear in the last 3minutes, and mammals in the last 39 seconds. In a fraction of the last second, our own species appears and all of human history and modern civilization occur. The presentation demonstrates very clearly the recency of our own arrival on Earth and the brevity of our tenure.

This exercise also demonstrates the link between changes in Earth's environments and the nature of life on Earth. For example, as oceanic life forms that gave off oxygen evolved, the gas accumulated in the atmosphere. After land plants appeared on Earth, amphibians came ashore and a long line of land-living animals evolved, including reptiles, dinosaurs, and mammals.

It is useful to close this exercise with one or more examples of events that illustrate the magnitude of humanity's impact on Earth's environments. Two such examples are the Industrial Revolution and development of the nuclear bomb. Another would be the invention of the automobile. It is particularly valuable to remind students that one of our greatest impacts on the environment is the transformation of land for agriculture and urbanization. You might end the class by showing an aerial photo (slide) of Earth's surface in a place clearly dominated by human activity, such as an industrial city or an area of intensive farming.

Reconstructing the Dates of Past Earthquakes

(These suggestions are also relevant to the section on stratigraphic sections and the clues they contain.)

Many of the review questions, thought questions, and exercises in the textbook can be used for in-class discussions. In fact, it is valuable for instructors to take the time to present some of these examples in class, because students will acquire a much better understanding of the course material if they have a chance to work some of the problems with an expert.

The following question is posed as Exercise 2 at the end of Chapter 3 and corresponds to Slides 3.1 and 3.2 in the Slide Set.

Along the coast of northern California, tidal platforms cut by wave action have been raised several meters out of the water during past earthquakes, and some are now covered with sand dunes. During earthquakes, all exposed, immobile marine life is killed instantly or dies soon after emergence above water. At one location, samples of fossil clam shells found in their original growth position on a platform contained only one-quarter the amount of ^{14}C in the modern atmosphere. How long ago was the earthquake that raised the tidal platform and killed the clams? A burned log left in a Native American fire pit was found in the bottom layer of dune sand covering the platform and shell. It contained half the amount of ^{14}C in the modern atmosphere. When did Native Americans camp on the dune that migrated atop the newly exposed platform? Do the radiometric ages make sense if you compare them with the sequence of events that you would have reconstructed using stratigraphic principles such as Steno's law of superposition? Why or why not?

The text that accompanies Slide 3.1 can be used to explain the stratigraphic section for this problem. It is helpful to sketch the section on the board and show its position relative to sea level. The emergent platform is about 3 m above sea level, and a modern wave-cut platform exists just below present sea level. Steno's principles of original horizontality and faunal/floral succession can be used to interpret the sequence of events one can glean from the section. For example, each of the sedimentary layers is horizontal, and the evidence of human occupation occurs at the top of the section.

If the clam shells found in growth position on the emergent platform contain only one-quarter the amount of ^{14}C in the modern atmosphere, and we assume that the atmosphere during the clam's lifetime contained a similar amount of ^{14}C as the modern atmosphere, then the time of the clam's death—which we will assume here was caused by coseismic uplift and hence is the time of a paleoearthquake—can be determined as follows.

One half-life of ^{14}C = 5730 years

After one half-life, only half the original carbon is left

After two half-lives, only one-quarter the original carbon is left

Two half-lives × 5730 years = 11,460 years

Likewise, the age of the log in the Native American fire pit is equal to one half-life, or 5730 years. Note that the actual radiometric ages given in the text with the Slide Set differ from those used in the textbook exercise. The exercise was designed to be a relatively easy example for students, not to correspond exactly to the real situation. The instructor can choose to tell students the real dates or simply say that the example in the slides is similar to that of the exercise.

We developed this example to be an environmental geology equivalent to the common practice of giving students in physical geology a stratigraphic section that shows volcanic intrusions, tilted and faulted sedimentary rocks, and so forth. Although this example is not as complex as those conventional examples, it does enable students to see how radiometric dating is used, and other examples presented in Chapter 3 are more complex. For example, you might wish to work with students to reconstruct the sequence of geologic events at Yucca Mountain, as presented in the stratigraphic section of Box 3.4 in the textbook.

TABLE I3.1 Time Analogy for Major Events in Earth's History

Event	Approximate Age (years)	Minutes and seconds of 50	Minutes of 50 (decimal)
Origin of Earth	4,600,000,000	0.00	0, 00
Hydrosphere and atmosphere form	4,200,000,000	4.35	4, 21
Oldest rocks that exist today form	3,960,000,000	6.96	6, 57
Single-celled bacteria evolve	3,500,000,000	11.96	11, 57
Algae and photosynthetic bacteria evolve	3,000,000,000	17.39	17, 23
First ice age; large continents develop	2,250,000,000	25.54	25, 33
Oxygen accumulates in atmosphere; multicelled life appears in oceans	1,800,000,000	30.43	30, 26
Oxygen-dependent life evolves in oceans	1,000,000,000	39.13	39, 08
Second ice age	800,000,000	41.30	41, 18
Proliferation of multicelled life forms in oceans	600,000,000	43.48	43, 29
Age of Trilobites	570,000,000	43.80	43, 48
Age of Corals; first land plants; third ice age	430,000,000	45.33	45, 20
Age of Fishes; first forests	400,000,000	45.65	45, 39
Age of Amphibians; first major coal-forming environments; fourth ice age	360,000,000	46.09	46, 05
Age of Reptiles	280,000,000	46.96	46, 57
Formation of Pangaea; dinosaurs appear	245,000,000	47.34	47, 20
Age of Dinosaurs; mammals appear; Pangaea breaks up	200,000,000	47.83	47, 50
India splits apart from Antarctica and moves northward; primates and flowering plants appear	120,000,000	48.70	48, 42
Age of Mammals; first true birds	60,000,000	49.35	49, 21
Whales appear; India collides with Eurasia	25,000,000	49.73	49, 44
Alps and Himalayas form	15,000,000	49.84	49, 50
Evolution of hominids	4,000,000	49.96	49, 57
Fifth ice age; upright-walking hominids (e.g., Lucy)	3,000,000	49.97	49, 58
Appearance of *Homo sapiens neanderthalensis*	125,000	50.00	50, 00
Appearance of *Homo sapiens sapiens*	50,000	50.00	50, 00
Last full glacial (maximum) conditions	20,000	50.00	50, 00
Agricultural Revolution	10,000	50.00	50, 00
Industrial Revolution	200	50.00	50, 00
First nuclear explosion	53	50.00	50, 00

DEMONSTRATIONS AND CASE STUDIES

Historical Methods of Estimating the Age of the Earth

Before the twentieth century, when radiometric dating techniques were developed that allowed us to pinpoint the age of the Earth accurately, scientists attempted to estimate Earth's age. The first to do so was Isaac Newton (1647–1727), the genius physicist who developed the laws of mechanics and put forth ideas about an orderly universe (see Chapter 1).

Estimates Based on Physicochemical Principles

In 1687, Newton showed mathematically how Earth cooled from an originally molten, liquid state. He wondered how long it would take a molten globe of iron with the same diameter as the Earth to cool. He estimated a minimum of 50,000 years and then posed a caveat: "But I suspect that the duration of heat may, on account of some latent causes, increase in a yet less proportion than that of the diameter." As he suspected, cooling time is a function not only of planetary diameter but also of other "latent," or hidden, sources of heat in the Earth, as discussed below.

Georges Louis Leclerc, Comte de Buffon (1707–1788), had access to data on the temperature of the Earth measured at hot springs and mines. These data indicated that the Earth is losing heat from its warmer interior. Buffon investigated cooling by manufacturing iron spheres of various diameters in his foundry. He identified a linear relation between sphere diameter and cooling time: as the diameter of a sphere increased, it took a proportionately longer time to cool. He then extrapolated from his largest sphere to the diameter of the Earth to estimate how long it would take an initially white hot Earth to cool to its present temperature. The result was about 75,000 years. But Buffon's largest sphere had a diameter of only 5 inches, whereas the diameter of the Earth is 7926 miles. The extrapolation was so great as to seem ludicrous.

The person most remembered for his calculations of Earth's age based on cooling time is the physicist William Thomson (Lord Kelvin) (1824–1907). Kelvin was elected president of the Royal Society of London for six consecutive terms and was chief scientist in charge of laying the first transatlantic telegraph cable, for which he was knighted by Queen Victoria. He was so highly esteemed that few ventured to doubt him or his work. In 1862, when he used fundamental principles of physics to "prove" that the Earth is 20 to 400 million years old, his estimate was considered authoritative. Kelvin published his results a few years after Darwin's *On the Origin of Species*, and Kelvin's work was cited as evidence that Darwin, and geologists, must be wrong about how long it took for the rock and fossil record to have been created.

Geologists were quick to question Kelvin's estimate, but his work seemed irrefutable because of the sophistication of his mathematical proofs. A highly esteemed American geologist, T. C. Chamberlin (1843–1928), bravely attacked Kelvin's analysis: "The fascinating impressiveness of rigorous mathematical analysis, with its atmosphere of precision and elegance, should not blind us to the defects of the premises that condition the whole process. There is perhaps no beguilement more insidious and dangerous than an elaborate and elegant mathematical process built upon unfortified premises." (Chamberlin, 1899, p. 224). Chamberlin recognized that Kelvin's calculations were done correctly but that the data Kelvin used and the assumptions he made were so full of uncertainty that the entire basis of his effort was flawed.

Kelvin made several assumptions when estimating Earth's age: The Earth was originally at a high temperature and has been cooling ever since. Earth's outer surface has cooled the most, and the interior is hotter. The surface of the Earth has a temperature that remains approximately constant as heat is conducted from its interior to surrounding space. In other words, Kelvin's assumption was that *no internal source of heat exists in the Earth. As heat flows out of the Earth, its temperature is reduced, so that it cools with time.*

Today, it is known that Kelvin's assumption was wrong. A source of internal heat, the radioactive decay of unstable atoms, was discovered just before Kelvin's death. As Newton had suspected hundreds of years earlier, some latent causes do contribute to the Earth's heat budget.

Estimates Based on Sodium in the Oceans

An Irish contemporary of Kelvin's, physicist John Joly (1857–1933), devised a clever way to obtain an independent estimate of Earth's age. His method examined the accumulation of sodium in ocean water as a result of the chemical weathering and erosion of land masses. Joly's model is similar to box models commonly used in environmental science and system dynamics today (see Chapter 2).

Joly viewed the ocean as a reservoir into which all rivers flow, carrying sodium—a common element in minerals—dissolved from rocks and washed into the streams. Joly reasoned that if he could determine the rate of addition of sodium to the ocean, and the amount of sodium presently in the ocean, he could calculate when the addition of sodium began (Figure I3.1). He thought that this time would be the same as when Earth cooled to a solid state and the ocean formed.

From the work of oceanographers, Joly knew the mean depth, area, and density of the ocean basins, from which he calculated the mass of ocean water as 1.3×10^{18} metric tons (mass = volume × density; volume = depth × area). The total salinity of the ocean is 3.5 percent, and sodium chloride ($NaCl$; common table salt) contributes 77.8 percent of this salinity. About 39 percent of the molecular weight of the $NaCl$ molecule is sodium, so Joly multiplied 39 percent × 77.8 percent × 3.5 percent × 1.3×10^{18} metric tons to get the total amount of sodium in the ocean, 1.4×10^{16} metric tons. He was also able to find published values of the volume of river water added to the ocean each year ($2.7 \times 10^4 \text{ km}^3$) and the sodium content of 19 of the largest rivers in the world (5250 metric tons of Na per km^3 of water). By multiplying these two numbers, he calculated that 1.4×10^8 metric tons of sodium are added to the ocean each year. Joly now had both of the numbers needed to determine the "age" of the oceans: the amount of sodium in the oceans today, and the rate at which sodium is added to the oceans each year. With these numbers, he could determine when the reservoir started to accumulate sodium. His answer was 100 million years ago.

But Joly, like Kelvin, made an incorrect assumption. Joly did not recognize that it is possible for sodium entering the ocean in river water to leave the ocean, rather than accumulating over time to make the ocean saltier and saltier. In fact, nearly half of the sodium that enters the ocean basins from the continents goes right back into the atmosphere as sea spray and moves along in the hydrologic cycle. The other half is precipitated as sedimentary mineral deposits on the ocean floor.

If half the sodium returns to continents as sea spray and the other half is deposited as sediments on the seafloor, then the concentration of sodium in the oceans must always remain the same. In fact, the concentration of seawater has been in approximate chemical equilibrium since early in Earth's history. When a reservoir is in a steady state, a state of no change over time, calculating the ratio of the total amount of something in the reservoir to the difference between inflow and outflow results in a residence time, as described in Chapter 2. Joly had determined the approximate residence time of sodium atoms in the oceans: the amount of time that a given sodium atom stays in the ocean, on average, before it is evaporated or deposited. Modern estimates of the residence time of sodium are very similar to Joly's values. Unfortunately, Joly did not determine the time at which inflow of sodium to the oceans began, and hence he did not determine the age of the Earth.

Lithosphere:
The Rock and Sediment System

CHAPTER OBJECTIVES

All the Earth system chapters (Parts II and III, Chapters 4 through 10), are organized in the same way. Each begins with basic materials and their origin and then addresses resource issues related to that system, hazards, and finally conservation or mitigation issues. In some cases, however, we had to deviate from this organizational scheme. One such case was the chapter on the lithosphere, which became so long in the early drafts that we finally had to divide it into two chapters. Those chapters are now Chapters 4 and 5. In Chapter 4, we focus on the materials and processes of the lithosphere, including minerals, rocks, plate tectonics, and faulting. In Chapter 5, we focus on resources, hazards, and mitigation of risks associated with hazards.

The material treated in Chapter 4 is typically presented in up to five or six chapters in physical geology textbooks. For example, a physical geology textbook might have one chapter on minerals, a separate chapter for each of the three rock types, a chapter on volcanic eruptions, and a chapter on earthquakes. In many books, discussion of mineral resources is saved for a final chapter.

We deliberated long and hard about what material is essential to an understanding of environmental geology, and that is what we present in Chapters 4 and 5. Although our textbook has only two chapters devoted to this material, we do not consider these chapters to be weak in physical geology. Instead, they present the most fundamental and essential information necessary to grapple with such issues as the environmental impact of copper mining at Butte, Montana. As with every chapter in this textbook, we try to present all scientific information in the context of relevant, important, and timely environmental issues. We consider it important to discuss mineral resources early in the book, not as an afterthought in a final chapter.

As we constructed these chapters, we asked ourselves such questions as, Does a student taking only this one science class in college need to be able to identify a particular mineral? Does knowledge of index minerals in metamorphic rocks help an introductory student to understand a particular issue of societal importance? Often, our answer to ourselves (and keep in mind that one of the authors is trained as a metamorphic petrologist) was that much of what we have traditionally taught in physical geology *is* essential for students who will later become geologists or other types of scientists. But what a nonscience student needs to know, if he or she is to have a somewhat sophisticated understanding of issues in environmental geology, is much more condensed and focused. At the same time, however, we do think that a grasp of the big picture is crucial. For these reasons, we've tried to present very specific information, such as the major mineral groups and rock types, in the context of types of plate boundaries and the interactions that occur along them.

With these specific and general goals in mind, the objectives of Chapter 4 are to

Investigate the nature and properties of Earth's minerals and rocks.

Identify some fundamental minerals and rocks.

Discuss the crustal and surface processes that produce minerals, rocks, and sediments.

Focus on the role of plate tectonics in altering Earth's surface over time.

CHAPTER OUTLINE

1. Lithosphere Materials
   A. Crystallization and Mineral Structure
   B. Major Mineral Groups
   C. Mineral Properties
   D. Major Rock Groups and the Rock Cycle

2. Plate Tectonics and the Rock Cycle
   A. Types of Plate Boundaries and Plate Interactions
   B. Divergent Plate Boundaries, Spreading Ridges, and Normal Faults
   C. Convergent Plate Boundaries, Subduction Zones, and Reverse Faults
   D. Transform Plate Boundaries and Strike-Slip Faults

3. Distribution of Rock Types
   A. Igneous Rocks
   B. Metamorphic Rocks
   C. Sedimentary Rocks

SUGGESTED LECTURE OUTLINE

The headings listed below are related to topics addressed in Chapter 4 and provide an alternative structure by which to present the material during classes and/or discussion periods. Items marked with one asterisk are treated in the following section on suggested lecture and discussion topics; those marked with two asterisks are treated in the section on demonstrations and case studies.

Materials that Compose the Lithosphere*
      Elements, minerals, and rocks
      Crystallization from melts and solutions
      Rock-forming minerals
      Using physical properties to identify minerals*
      Mineral classification
            Major mineral groups
            Silicate mineral structures and hazardous minerals (asbestos)
      Major rock groups

Plate Tectonics
      What drives the rock cycle?
      Types of plate boundaries
      Plate interactions and faulting

Conceptual and Systems Views of the Rock Cycle
      Magma types and plate boundaries
      Rock reservoirs and fluxes*
      Rock types and their identification**

SUGGESTED LECTURE AND DISCUSSION TOPICS

Materials That Compose the Lithosphere

Students benefit from the instructor's help in making connections between content presented in different chapters of their textbooks or in different lectures. Chapter 4 focuses on the lithosphere, which was introduced briefly in Chapter 2 during a discussion of all the basic systems, or spheres, at and near the Earth's surface. The instructor can remind the student of this background material on the lithosphere in Chapter 2, on pages 41–42. In addition, the development of the lithosphere during Earth's early physical differentiation is presented on pages 38–39, and Figure 2.8 illustrates Earth's layered configuration.

Using Physical Properties to Identify Minerals

In 1997, we tried something new in our introductory classes and found it to be very successful. During the lecture portion of the course, before the labs on minerals and rocks, we presented students with samples of several common minerals but did not reveal their names. We used the following minerals:

> Sulfur
> Halite
> Hematite
> Graphite
> Talc

We got small samples of these minerals from *Ward Scientific Supply*. The samples come 20 in a box and are very inexpensive ($5 to $10 per box). We placed the samples at the students' desks before class, giving one of each of the five samples to every two students. We also placed a nail in the pile of minerals.

When the students arrived, we asked them to use all their senses to try to distinguish the minerals from one another and to identify them by name if they could. We also asked them to list any of the common uses of each of these minerals.

For 10 minutes, students played with the samples and began animated discussions about what the minerals might be. We encouraged them to talk with other student pairs if they wished. They smelled the sulfur, and most recognized its odor. They wrote with the graphite on their papers (and bodies!), and a few tasted the halite. On their own, they noticed immediately the much greater density of the hematite. They quickly realized that the talc and graphite are very soft, and they began scratching all the samples with the nail. The students were exploring and discovering, and they clearly enjoyed the process.

As the activity began to die down, we asked the students to help make a list of the physical properties that enabled them to distinguish one mineral from another. With little prompting, the students came up with the following list, which we wrote on the blackboard. (This list could be done on an overhead projector as well.)

> Density (many students said "weight," and we explained the difference between weight and density)

> Hardness
> Smell
> Taste
> Color

We helped students to recognize yet another property, cleavage, and defined it for them. Then we passed around a large sample of mica and noted that mica, talc, and graphite all have the same cleavage—perfect in one direction—whereas halite has three perfect cleavage planes at right angles to one another, hematite has no cleavage, and sulfur has poor cleavage in two directions.

At this point, we explained that geologists use each of these physical properties to identify minerals, and we constructed a table of the five minerals and their physical properties on the board (see Appendix 4 in the textbook for mineral properties). We also gave the mineral formula for each and noted that one mineral is an oxide (hematite), two are native elements (sulfur and graphite), one is a silicate (talc), and the fifth is a chloride (halite). We commented on the fact that minerals are made of atoms of different elements and that a different formula results in a different mineral with different physical properties. For example, adding one more iron atom and one more oxygen atom to each molecule in hematite would result in magnetite, which is magnetic. We concluded the session with a summary of the uses of each of these minerals in industry, agriculture, and other human activities. Again, we called on the students for suggestions and augmented their contributions with our own.

Asking the students to determine for themselves how to identify the physical properties of minerals allows them to use scientific thinking and reasoning. This exercise got them engaged in a discussion of minerals and made it easier for us to proceed with the lecture material on minerals.

Rock Reservoirs and Fluxes

As noted, students benefit from faculty expertise to help them link material from different chapters. We think it is particularly important in an Earth system science approach to help students see the similarities between the conceptual model of the rock cycle presented in Chapter 4 (Figure 4.11) and the system dynamics view presented in Chapter 2 (Figure 2.19). Many students will have been exposed to the conceptual model in junior high school Earth science classes.

Whereas the conventional model of the rock cycle focuses on processes and rock types, the systems model emphasizes the sizes of reservoirs that contain rocks and sediments and the magnitude of fluxes among them. The systems view, in fact, is more closely aligned with the plate-tectonics cycle than is the conceptual view. It is very easy to move from a discussion of the systems model in Figure 2.19 to the relationship of different rock types to plate boundaries and their interactions. Figure 4.20 is a good illustration to use for the relationship of igneous rock types to plate boundaries. The instructor can add information about sedimentary and metamorphic rocks to this diagram by discussing sediment in trenches or high-pressure/high-temperature metamorphism along subduction zones.

DEMONSTRATIONS AND CASE STUDIES

Identifying Rocks in the Field

Students need help in learning how to identify rocks other than the samples they see in drawers in the laboratory. The major clues to identifying a rock in hand specimen are its mineralogy and texture. In the field, however, geologists typically look at other characteristics as well, such as the relationship of the rock to those around it and the presence of fossils in an outcrop. We will examine such features in combination with mineralogy and texture to illustrate how to use the accompanying flowchart (Figure I4.1) to identify some common examples of igneous, sedimentary, and metamorphic rocks.

This exercise can be done in class or as a group in the lab with guidance from the instructor just before students work on their own or in teams. The instructor can display a slide of each rock type and show the students how to use the flowchart to identify the rock. Instructors can also use any of their own slides. This exercise is a useful alternative to the traditional labs on sedimentary, igneous, and metamorphic rocks.

The first sample (Figure I4.2) was taken from an outcrop in Hawaii that contains fossils of coral, so it must be sedimentary. From the flowchart, one would conclude that the rock is clastic, because it contains much broken rock material. It fizzes if sprinkled with acid, so it is a limestone. Specifically, it is a fossiliferous limestone, because some fossils are visible.

The second sample (Figure I4.3) was taken from an outcrop in Australia. The sample is clearly crystalline. Crystals of pink feldspar (orthoclase; some with white feldspar, or plagioclase, rims), white feldspar (plagioclase), and quartz are nearly as large as the Australian coin shown in the slide. Surrounding these crystals are much smaller biotite crystals. The crystals have interlocking boundaries, as in a mosaic pattern, indicating that they cooled from a melt. According to the flowchart, if a crystalline rock does not fizz—and this one does not—the next step is to determine the texture of the crystals. In this sample, all crystals of several minerals are visible, and the crystals are white, pink, and gray in color. Thus, the sample is a granite. This granite is actually the Kameruka granodiorite from New South Wales. For the purposes of an introductory environmental geology class, however, it is sufficient to classify it as a granite.

The third sample (Figure I4.4) was taken from an outcrop in Newfoundland. Although bedding planes (dipping to the left at a high angle) are prominent, so too are foliation planes (vertical). It is important for the students to see both features in the same rock and to recognize that the foliation is an overprint on the original rock fabric. According to the flowchart, if a rock is foliated, the next step is to examine the arrangement of minerals in the rock. Although this is somewhat challenging to do without an actual hand specimen, it is clear from the slide that the rock has sheety minerals that are aligned with one another, and the instructor can inform the students that the minerals are coarse-grained, making this rock a schist.

CHAPTER 5

Lithosphere: Resources, Hazards, and Change

CHAPTER OBJECTIVES

Chapter 5, a companion and follow-up to Chapter 4, focuses on resources and hazards of the lithosphere. The resources treated are ore minerals, and the hazards discussed are volcanism and earthquakes. Included in the discussion of ore minerals are the geologic processes that concentrate the resources in localized areas and the environmental impact of mining. For volcanic eruptions and earthquakes, we follow a similar organization: processes and causes, assessment of hazards, and minimization of risk. The stated goals of the chapter are to

Examine the origins of ore minerals and the issues connected with their extraction.

Examine the characteristics of volcanoes and ways to predict eruptions.

Investigate the causes of earthquakes and ways to minimize the associated risks.

The title of the chapter includes changes in the lithosphere. In an early draft of the textbook, each of the systems chapters (Chapters 4 through 10) contained a final section on environmental and global change. To shorten the chapters, however, we condensed those sections into boxes (see the list on page xiii in the textbook) and placed some of the remaining material in appropriate places throughout each chapter. In this chapter, for example, we discuss the fact that the processes that produce some ore minerals no longer occur on Earth, such as the eruption of nickel-bearing ultramafic volcanic magmas that became komatiites (see page 128 in the textbook).

This chapter also contains an example of the cycling of matter. Explaining the ways in which copper and other metals are cycled from one Earth system to another makes it relatively easy to demonstrate the magnitude of the environmental impact of mining. In the cases of mercury and lead, for example, atmospheric emissions are now several hundred times greater than natural background rates. The example of copper cycling is used to demonstrate how to calculate residence times of copper in various Earth systems. The utility of this approach is shown by using the residence times to assess the environmental impact of copper smelting on soil and air in Tennessee (see pages 137–138 in the textbook).

CHAPTER OUTLINE

1. Rock and Mineral Resources
    A. Ore Deposits and Ore Minerals
    B. Types of Ore Deposits
    C. Depletion of Mineral Resources
    D. Environmental Impact of Mining

2. Volcanic Eruptions
    A. Magma Types, Eruptive Styles, and Volcanic Landforms
    B. Shield Volcanoes, Stratovolcanoes, and Calderas
    C. Assessing Volcanic Hazards
    D. Minimizing the Risks of Volcanism

3. Earthquakes
    A. The Causes of Earthquakes
    B. Assessing Earthquake Hazards
    C. Minimizing the Risks of Earthquakes

SUGGESTED LECTURE OUTLINE

The headings listed below are related to topics addressed in Chapter 5 and provide an alternative structure by which to present the material during classes and/or discussion periods. Items marked with one asterisk are treated in the following section on suggested lecture and discussion topics; those marked with two asterisks are treated in the section on demonstrations and case studies.

Human Consumption of Mineral Resources
    Minerals and human history: from Stone Age to Iron Age—or is it New Stone Age?*
    Mineral consumption in industrialized nations
    Economical concentration factors and the problem of waste rocks

The Origin of Ore Minerals and Deposits
    Hydrothermal ore deposits and plate tectonics
    Igneous ore deposits
    Sedimentary ore deposits
    Cycling of metals through Earth systems

Problems Associated with Mineral Resources
    Depletion of mineral resources—is it a threat?
    Reserves and resources
    Environmental impact of mining*
    New environmental regulations for the mining industry

Volcanoes and Volcanic Eruptions
    Shield volcanoes, stratovolcanoes, and calderas
    Recent volcanic eruptions and their societal impact*
    Predicting volcanic eruptions
    Volcanic hazards maps

Earthquakes
    Elastic rebound and seismic energy release**
    Seismic waves**
    Earthquakes along plate boundaries
    Earthquakes in the interiors of tectonic plates**
    Predicting earthquakes**

SUGGESTED LECTURE AND DISCUSSION TOPICS

Minerals and Human History: From Stone Age to Iron Age—or Is It New Stone Age?

Typically, students are surprised when told the amounts of mineral resources used by people in industrialized nations. To raise awareness of their dependence on large supplies of inexpensive minerals, the instructor can begin by asking the students what they have used that day that is made of minerals. With some thought, most students will be able to name a long list of items. Here are typical responses to this question:

    Silverware, knives
    Automobiles, buses, subways, and so forth
    Soda cans (one of the largest consumers of aluminum)
    Bicycles
    Electrical wiring
    Paper clips
    Aluminum foil
    Coins (money)
    Pens
    Water pipes
    Sewage pipes
    Metal tubs and sinks

We bring samples of metals and glass into class, including a soda can, wound copper wire, a piece of steel, a piece of rusted iron (such as a nail), and a Pyrex beaker. The wound copper wire can be used to demonstrate the value of copper: it is ductile yet strong and is a good conductor of electricity. The soda can is light yet strong, demonstrating the value of aluminum, which can be rolled into extremely thin sheets.

Aluminum is an interesting metal because it is found in only one type of ore deposit: bauxite ores formed from prolonged weathering (the process is described in Chapter 6). The Pyrex beaker is made from quartz, but the manufacturer added sodium to the mix to make the beaker heat resistant. Before class, the instructor can fill the beaker with sand to demonstrate the origin of the glass beaker (see also Figure 5.1 in the textbook). Like the Pyrex beaker, the piece of steel is a hybrid, in this case a mix of iron and carbon. The mix makes the steel stronger and less brittle than plain iron.

The instructor can quote some of the values given in the textbook about how much of each mineral is used per person per year in the United States, but even more dramatic is to quote how much each new-born American will need over a lifetime (see page 125 in the textbook). Most students are shocked to learn that they will use 16 tons of iron or 14 tons of salt. Remind the students, however, that much of these minerals is used by industry and agriculture, and the values are merely averages obtained by dividing U.S. population into total national mineral consumption.

The changes in human civilization that have occurred as people developed the technology to use different mineral resources are of historical interest. The textbook briefly describes some of these changes, beginning on page 125. The Bronze Age was very short compared to the Iron Age, which continues today. More iron is used, largely for steel production, than any other metal. Modern industry also relies on large volumes of clay and glass (from quartz), so some scholars suggest that the present is a time of shifting from the Iron Age to the New Stone Age.

The Environmental Impact of Mining

Chapter 5 opens with a photo of the Grasberg copper-gold mine in Irian Jaya, on the island of New Guinea. A recent series of three articles in *Geotimes* magazine highlighted the geology, environmental impact, and investigative aspects of this mine. Written by Mark Cloos, who has worked in the region for nearly a decade, the articles are short (3–4 pages each) and clearly written. Of the three articles, the first two are suitable for reading assignments for introductory geology students.

The first article in the Grasberg series, "The Anatomy of a Mine," was published in January 1997. It describes the discovery and development of the copper-gold ore body and gives the results of tests to determine the amount of copper and gold in it. The instructor can ask students to read this article and find out how the ore body was discovered and how to use concentration factors to calculate how much ore must be mined to get the ore minerals.

The second article, "Responsible Mining: Environmental Management at Grasberg," was published in May 1997. It focuses on environmental management at the mine site. If students are given this second article to read, the instructor can ask them to make a list of the types of environmental impact resulting from mining and the solutions developed by Freeport-McMoRan Copper and Gold and its Indonesian subsidiary, PT Freeport Indonesia, to minimize negative impacts. As described carefully in the article, the greatest impact on the environment is associated with mine waste rock, which is crushed and released into local rivers.

Recent Volcanic Eruptions and Their Societal Impact

Geology instructors are fortunate in that a volcano is always erupting somewhere, and today it is possible to bring that eruption into the classroom via the Internet. At the time of the final writing and printing of *Environmental Geology* (the end of 1997), the world's best known volcanic activity was occurring on a tiny island in the Caribbean called Montserrat. The eruptions began in the Soufriére Hills in July 1995, and since then about half the island's population—some 5000 people—have been evacuated. Many moved to the northern part of the island, where older volcanic centers have long been inactive. Others moved to England or other islands in the Caribbean.

Because textbooks do not contain information about events that occur after the book's printing, instructors have the opportunity to relate the example of a current event to the material in the text. A number of web sites contain photos, and even video footage, of the numerous explosions of gas, ash, and rock that have occurred at Montserrat since 1995 (see the W. H. Freeman web site on Montserrat, http:/www.whfreeman.com/montserrat). During class, the instructor can bring up images of an eruption and explain its tectonic setting. The Soufriére Hills, for example, are located along a volcanic arc, the Lesser Antilles, that formed where the South American Plate plunges beneath the more buoyant and less dense Caribbean Plate. Sediment and water scraped down into the subduction zone has contributed to formation of a magma that is mafic to andesitic in character. As magma rises upward from the subducting slab and offscraped sediment, it is accumulating in a dome called New Dome, amid several others that were active hundreds to tens of thousands of years ago.

DEMONSTRATIONS AND CASE STUDIES

The Earthquake Machine: A Physical Model

A physical model (Figure I5.1) using a brick, sandpaper, a bungee cord, and a crank can be used to demonstrate many of the fundamental aspects of earthquake behavior. The brick, attached to the crank with a bungee cord, sits on a board covered with sandpaper. Initially, the brick is at rest. As the crank is turned, shear stress develops along the plane between the brick and the sandpaper. This plane is analogous to a fault plane, and the brick and sandpaper are analogous to blocks of rock on each side of a fault. Because of frictional resistance between the brick and the sandpaper, the brick does not move for some time. Rather, the bungee cord gets tauter and tauter, storing strain in the form of elastic energy. The more energy stored, the greater the amount that can be released when the brick finally does slip. Cranking the earthquake machine is analogous to the continuous motion of the Earth's upper mantle because of convection, and the bungee cord is similar to the elastic crust, which can store seismic energy.

At some point, the brick might slide just a tiny bit, making a small grating sound and releasing some of the stored energy. Then, suddenly, the brick might slide several inches or more. The small event is analogous to an earthquake foreshock, which is caused by minor amounts of slip along a fault plane just before the much larger amount of slip that occurs during the main shock of a large earthquake. After the main shock, small amounts of slip might again occur along the fault plane, just as aftershocks commonly follow large earthquakes. This sequence of events is similar to that observed in real earthquakes over the earthquake cycle: accumulation and release of strain in crustal rocks is repeated over and over, with long periods of seismic quiescence (little or no activity) punctuated by sudden events that release much of the pent-up strain energy (Figure I5.2).

In the earthquake machine, even if the crank is turned slowly and steadily, the recurrence of slip events can be irregular. Sometimes the brick doesn't move for a while, then it slips a large amount; at other times, it slips lesser amounts but more frequently. This happens with earthquakes, too. The time that elapses between events is the recurrence interval, and the average amount of slip per unit of time is the slip rate. In the Earth's crust, slip along faults sometimes is minimal, causing small earthquakes; at other times it is substantial, causing very large earthquakes. Actual recurrence intervals and slip rates vary substantially, with some areas characterized by rare but large earthquakes and others by frequent but smaller earthquakes. In general, where the rates of plate motion are the greatest, the amount of earthquake activity and the rates of slip are the highest and the duration of time between events is the shortest.

The energy released during an earthquake is proportional to the size (length and width) of the fault plane that ruptures and the volume of rock deformed along the fault. For example, one cubic meter of rock can release about 10^9 ergs of seismic energy during an earthquake. This is about as much energy as in a firecracker. If only a few cubic meters of rock were to slip, the energy released would be small and only the most sensitive instruments could detect the event. When larger amounts of rock slip, a great deal of seismic energy can be released. For example, if the fault rupture plane is about 400 km long and 10 km deep, and rocks as far away as 50 km on each side of the fault are distorted, then the total volume of rock that releases elastic strain energy is $(400 \times 10^3 \text{ m}) \times (10 \times 10^3 \text{ m}) \times (100 \times 10^3 \text{ m})$, or $4 \times 10^{14} \text{ m}^3$ of rock. The total amount of strain energy released is 4×10^{23} ergs. These dimensions are similar to those of the 1906 San Francisco earthquake (see Figure I5.2).

Seismic Waves and Travel Times: An Example

On October 17, 1989, at about 5:15 P.M., a student at Franklin and Marshall College in Lancaster, Pennsylvania, called one of the authors of this book to report that a large earthquake had just begun to register on the seismograph. The ground beneath the college was shaking a tiny, imperceptible amount, and although the local inhabitants—including the student—felt nothing, the seismograph showed that a large earthquake had occurred somewhere. Even the largest earthquakes last only about one to two minutes, so the actual event probably was over already. By the time the dampened seismic waves reached and passed through the ground beneath the college, much of the earthquake's stored energy had been converted to heat by frictional resistance. With increasing distance from the earthquake source, seismic wave amplitudes become smaller and smaller, until only the most sensitive instruments can detect their arrival.

The professor explained that some seismic waves travel faster than others and so would arrive first. The amount of time that elapsed before a second type of seismic wave arrived could be used to determine the distance between the earthquake source and the seismometer. The principle is similar to the one used to determine how far away lightning strikes, based on the elapsed time between the arrival of faster moving light waves and slower moving sound waves (thunder). The farther away the lightning strike or earthquake source, the greater the time that elapses between the arrival of the two different wave types. The student making the call examined the seismogram and detected two prominent batches of wave arrivals. He reported that the elapsed time between the arrivals was about five minutes. The professor, who knew the rates at which the two wave types travel, concluded that the earthquake occurred about 3000 km from the college.

At a distance of 3000 km, the earthquake must have begun about 8 minutes before the ground beneath the college began to shake, because the fastest seismic waves move at about 6 km/s. The professor immediately turned on a television set, assuming that 20 minutes was long enough to result in media coverage. If there were no mention of an earthquake, the event might have occurred to the east, in the middle of the Atlantic Ocean, thus harming no one. Unfortunately, however, several stations were reporting a large earthquake to the west, in the San Francisco area, as well as the disruption of the World Series baseball game which had begun at Oakland stadium just before the earthquake struck. At 5:30 P.M., stunned spectators were still filing from the stadium.

Assessing Earthquake Hazards: Intraplate Earthquakes

Hazardous earthquakes are not limited to zones of plate collision. Recent work in the central United States, far from any active plate boundary, has determined that the great earthquakes that occurred near New Madrid, Missouri, in 1811–1812 pushed a hill into the path of a stream flowing into the Mississippi River. The blocked stream formed a new lake, now known as Reelfoot Lake, and inundated bald cypress trees. These trees still exist today, but they do not germinate underwater. Cores from these trees indicate that the submergence resulted in dramatically increased growth rates.

Like other intraplate earthquakes that occur in the interiors of tectonic plates, the New Madrid earthquakes took place along faults associated with an ancient plate boundary. In this case, the ancient boundary is the Reelfoot Rift, a failed rift underlying the Mississippi River basin (Figure 15.3). Maps of recent earthquakes, even those that are too small to be felt by humans, help to define the fault planes of these ancient structures.

Although intraplate earthquakes occur less often than earthquakes at plate boundaries, a number have occurred in recent decades throughout the world. An intraplate earthquake in Latur, India, killed more than 11,000 people in 1993. Intraplate events account for only about 0.5 percent of the global release of seismic energy each year, yet they are some of the most hazardous earthquakes. Inhabitants of the ancient bedrock interiors of continents, euphemistically called stable platforms, are far less prepared for earthquakes than people living in zones of frequent seismic activity. Furthermore, seismic energy can move rapidly through the dense, older crust of continental interiors; thus, the effects of a single earthquake are felt over much greater areas than those of an earthquake occurring in shattered, young continental margin rocks along a plate boundary.

Assessing Earthquake Hazards: GPS Measurements

A number of earthquakes in the western United States in the 1980s and 1990s shifted attention from mapped and obvious faults, such as the San Andreas, to smaller and often unknown faults, such as the one that caused the Northridge earthquake (magnitude 6.8) near Los Angeles in 1994. The Northridge event, which caused 57 deaths and injured nearly 10,000 people, occurred on a blind thrust fault—a fault that is buried and does not reach the surface.

Although a scarp did not form at the ground surface during the Northridge earthquake, Global Positioning Satellite (GPS) measurements of the positions of ground stations before and after the event indicate that the ground moved horizontally by more than 20 cm and vertically by more than 30 cm during the earthquake. GPS measurements provide the strain information needed to assess the potential size and frequency of future earthquakes. Such an assessment was recently completed for the entire Los Angeles basin, which contains at least a dozen earthquake-producing faults other than the San Andreas within the heavily urbanized area. Although no one of these faults is likely to produce an earthquake as large as one occurring on the San Andreas fault, all of them are located immediately beneath urban areas, posing as much hazard locally as the more distant San Andreas fault.

Even if an earthquake has not occurred in a region in historic times, clues to such past events, or paleoearthquakes, often exist. One such clue might be peats buried by sand when the ground gives way. Paleoseismology, the study of these clues, allows geologists to interpret the history of earthquake activity along a specific fault. Layers of sediment commonly provide the most clues. In particular, organic matter in sediments can be radiometrically dated based on decay of the radioactive carbon-14 isotope (see Chapter 3). Dated layers can then be used to bracket the timing of past earthquakes. Paleoseismology is especially useful because the historic record of earthquakes is so limited (usually less than 200 years), and large earthquakes typically recur at intervals of centuries.

Along the south central San Andreas fault, geologists have dated charcoal buried in peat, sand, and mud, which has allowed them to reconstruct a history of large events that spans 1400 years. One earthquake occurred about the time that Christopher Columbus first reached the West Indies. Studies similar to this one throughout the world indicate that recurrence intervals between earthquakes are irregular, and some evidence suggests that earthquakes along specific fault segments tend to cluster in time. For this reason, it is not likely that estimates of average recurrence intervals can ever be used as a reliable tool for predicting the exact timing of future earthquakes. Nevertheless, paleoseismology provides valuable evidence about the likely frequency and magnitude of earthquakes, both of which are important criteria for assessing the probability of events of a certain size occurring over a given period of time.

Soil Systems and Weathering

CHAPTER OBJECTIVES

Chapter 6 focuses on one of Earth's major systems, the pedosphere. It presents material on the weathering and soil-forming processes that create the pedosphere; resources produced from weathering, including fertile soil, clay, laterite, and mineral ores; hazards associated with soil and weathered debris, including soil erosion, desertification, and mass movement; and the mitigation of these hazards. The stated goals of the chapter are to

Examine the processes of weathering and soil formation that created the pedosphere and the resources in it that we use.

Explore the causes of the hazards of soil erosion, desertification, and mass movement, and consider how they can be minimized.

Investigate examples of changes in the pedosphere over human and geologic time periods.

Examples of global change and cycling of matter are also contained in this chapter. Box 6.1 explains how mountain building might lead to greater rates of chemical weathering and hence a drawdown in the amount of carbon dioxide in the atmosphere, which in turn might cause climate change (global cooling). The nitrogen cycle is explained in detail on pages 173–174 and is illustrated in Figure 6.13. Human impacts on this cycle through farming and fertilizer production are discussed in particular.

CHAPTER OUTLINE

1.   The Pedosphere: A Geomembrane to Other Earth Systems
      A.   Physical Weathering
      B.   Chemical Weathering
      C.   Soil Profiles and Soil-Forming Factors
      D.   The Interaction of Earth Systems to Form Soil

2.   Pedosphere Resources
      A.   The Fertile Soil
      B.   Clay, Laterite, and Mineral Ores

3.   Soil Erosion Hazards and Soil Conservation
      A.   Soil Erosion
      B.   Assessing Soil Erosion
      C.   Soil Conservation Practices

4.   Mass Movement Hazards and Their Mitigation
      A.   The Roles of Gravity and Water in Mass Movement
      B.   Types of Mass Movement
      C.   Causes and Prevention of Mass Movement

SUGGESTED LECTURE OUTLINE

The headings listed below are related to topics addressed in Chapter 6 and provide an alternative structure by which to present the material during classes and/or discussion periods. Items marked with one asterisk are treated in the following section on suggested lecture and discussion topics; those marked with two asterisks are treated in the section on demonstrations and case studies.

Weathering of the Earth's Crust
      Physical weathering
            The importance of worms in physical weathering**
      Chemical weathering
            Rates of weathering and their impact on global climate
            The formation of resources from weathering (bauxite and laterite)

Soils
      Soil-forming processes
      Soil horizons and profiles
      Soils and the carbon cycle*
      The colors of soil horizons, and what they tell us about soil formation**
      The roles of clay, sand, and silt in soils
      Soil-forming factors: climate, time, parent material, organic matter, and topography
      Soil fertility and soil maps
      Soil erosion: rates and measurement**
      Soil conservation practices

Mass Movement
    Gravity and the angle of repose**
    Mass movement classification
    Deforestation and mass movement
    Minimizing mass movement hazards

SUGGESTED LECTURE AND DISCUSSION TOPICS

The Effect of Carbon Storage in Soils

The carbon cycle is introduced in Chapter 10 in the context of ocean processes, but instructors might want to introduce it along with the nitrogen cycle in Chapter 6. Questions related to weathering and the carbon cycle that could be posed at the beginning of the lecture on soils include

How might rates of weathering affect global climate?

How might the increase in farming since the Agricultural Revolution affect global climate?

How might the melting of vast ice sheets in the past 10,000 years affect the amount of carbon stored in soils?

Once these questions have been posed, the instructor can present material from the text that follows to demonstrate that soils are an intimate part of Earth systems (hydrosphere, atmosphere, lithosphere, and biosphere) and that they continuously interact with one another and change in response to processes operating in each system.

Carbon is both stored in and released from the pedosphere, and the amounts of each flux vary over time. As a consequence, the pedosphere plays a role in the climatic changes associated with fluctuating amounts of carbon dioxide in the atmosphere. Part of the change in the pedosphere is a result of the advancement and retreat of ice sheets, which alternately cover the land and expose it to the atmosphere. Some of the more recent changes, however, might result from agricultural activities.

A substantial amount of the world's carbon is stored in the pedosphere in two main forms: as plant litter and as soil organic carbon. Soil organic carbon is organic matter in various states of decomposition (that is, humus). About 17 percent of the carbon in the world's carbon cycle is stored as litter and 44 percent as soil organic carbon (excluding carbon stored in rocks, the deep ocean, and fossil fuels; see Figure 10.12 in the textbook). Plant litter is a source of carbon input to the soil organic pool, and on a global scale the mean residence time of carbon in soil is very short, only 26 years. (The residence time is calculated by dividing the global soil organic matter by the global litterfall.) The residence time, however, varies with climate, which affects the rate at which organic matter decomposes.

For example, 18 percent of the total world soil organic carbon is stored in tropical forests, but the amount of soil organic matter per square meter of soil is less than that for boreal forests because rates of decomposition are slower in cooler climates. More of the total world carbon is stored in tropical forest soils because they cover a land area about twice that of boreal forests. Several other ecosystems, including temperate forests and grasslands and arctic tundra and alpine areas, contain fairly large pools of soil organic carbon because temperatures and rates of decomposition are low. With the exception of wetlands (swamp and marsh ecosystem), the tundra and alpine ecosystem contains the largest amount of mean soil organic matter per square meter.

If climate, environment, or ecosystem change, so will the amount of carbon stored in various parts of the world. Just before the interglacial warm period that began about 10,000 years ago, much of the northern hemisphere was covered with thick ice sheets, and much of the tundra that today contains so much soil organic carbon was frozen and buried. At present, climatic warming is causing rates of decomposition to increase in most ecosystems. Some climate modelers suggest that the uncovering of land during deglaciation about 10,000 years ago and the continued thawing of tundra during the Holocene warm period and at the present time are partly responsible for the observed increases in carbon dioxide in the atmosphere at those times.

The transformation of about 10 percent of the world's soil to agricultural land also is affecting the pool of carbon in soil organic matter and might be another cause of the increased amount of carbon dioxide in the atmosphere in the past few hundred years. Cultivated land has less litterfall and higher rates of decomposition than unaltered land. When native land is converted to agriculture, the amount of soil organic matter declines for these reasons.

If climate, environment, or ecosystems change, so will the amount of carbon stored in various parts of the world. Just before the interglacial warm period that began about 10,000 years ago, much of the northern hemisphere was covered with thick ice sheets, and much of the tundra that today contains so much soil organic carbon was frozen and buried. At present, climatic warming is causing rates of decomposition to increase in most ecosystems. Some climate modelers suggest that the uncovering of land during deglaciation about 10,000 years ago and the continued thawing of tundra during the Holocene warm period and at the present time are partly responsible for the observed increases in carbon dioxide in the atmosphere at those times.

DEMONSTRATIONS AND CASE STUDIES

Please see the previous section, Suggested Lecture and Discussion Topics, for an additional case study on the effect of carbon storage in soils.

Basic Soil Properties: Color and Texture

The color and texture of a soil provide qualitative evidence of the processes that produced it. In most soils, color and texture vary with depth, because the degree of addition, transformation, transfer, and removal of minerals and organic matter differs among soil horizons.

Dark colors in the A horizon indicate the accumulation of organic matter; light gray and whitish colors imply that the humus content is minimal, perhaps because of extensive leaching by subsurface water. On the other hand, light gray and white colors in the B horizon in a semiarid or arid region can signify the accumulation of salts. Water carrying dissolved ions can become supersaturated as a result of evapotranspiration. When this happens, calcium carbonate, gypsum, and some other salts can precipitate into the soil. In much of the American Southwest, desert soils contain nodules, lenses, and even thick layers of hard calcite, known as caliche, that has precipitated over thousands of years.

The oxidation of iron-bearing minerals colors the B and C soil horizons red; thus, red colors in these horizons indicate that the soil has good drainage and abundant moisture and is well aerated. The bright red soils common in Hawaii, the southeastern United States, Central and South America, and Southeast Asia owe their colors to iron oxides released by extensive weathering. In soils that are not as well drained, and in which oxygen is less likely to move through pore spaces, one might find the colors blue-gray (indicating the presence of iron compounds) or yellow (iron hydroxide minerals).

Texture refers to the proportion of particles of various sizes in the soil. Although soils can contain particles coarser than sand, soil texture classification refers to the particles that are sand-sized or finer. The texture of the inorganic (mineral) fraction of the soil is determined using a triangular diagram that represents the proportions of clay, silt, and sand (Figure I6.1). The clay-sized fraction includes true clay minerals as well as other crystalline minerals and amorphous (noncrystalline) substances that are clay-sized. The proportions of sand, silt, and clay can be determined in a laboratory using equipment designed to separate the fractions, but many soil surveyors and geoscientists learn to estimate the percentages in the field, simply by touch.

For example, a scientist may estimate that a soil contains 20–45 percent sand, 28–40 percent clay, and 15–52 percent silt; from Figure I6.1, the soil is a clay loam. Loam is one of the best soils for farming, because it contains roughly equal parts of silt and sand and less than 30 percent clay. When present in more than about 30–50 percent, clay contributes to poor drainage and prohibits movement of nutrients and water to plant roots. Some clay is desirable, because a soil composed of all sand and silt is less able to retain moisture and nutrients and is easily eroded. Although silt and sand are relatively inert and do not attract or repel ions, they contribute to soil permeability, drainage, and strength.

Measuring Accelerated Soil Erosion Rates

Soil erosion rates are traditionally determined by upstream, midstream, and downstream methods. Researchers who apply upstream methods use a combination of techniques to measure the actual amount of erosion at the site of soil loss. Midstream methods are based on estimates of the volume of sediment carried by rivers and streams, and researchers then use the estimates to calculate how much soil must have been removed from the slopes upstream. Downstream methods involve measuring the volume of sediment deposited in a lake, reservoir, or other basin and using this information to calculate upstream rates of erosion. Midstream and downstream methods rely on numerous assumptions and are always approximations: for example, some sediment might escape notice by being trapped in various places throughout the watershed, such as floodplains, and some sediment might be carried by winds rather than by water. Even upstream methods yield only very short-term rates if they are based on monitoring studies done over a period of only a few years.

Researchers in Kenya (one of whom was Tom Dunne of the University of California at Santa Barbara) have developed an ingenious method for estimating soil erosion rates that cover a period of tens of years: they measure how much the ground surface has dropped around the bases of trees (see Figure 6.18 in the textbook). Working in heavily grazed semiarid range land, on a slope of about 9 percent, the scientists found tree roots exposed by erosion and used the ages of the trees to estimate erosion rates. The trees were up to 40 years old, allowing estimates of soil erosion to be made from about 1936 onward. Average erosion rates from about 1936 to 1961 were 1.9 mm/year. After 1961, erosion rates increased to 5.5 mm/year. The natural rates of soil formation on undisturbed soils are about 0.02 to 0.11 mm/year, so these rates of erosion are far in excess of rates of soil renewal. The increase in erosion rates in the 1960s coincided with a period of catastrophic rainstorms following a decade of drought, and then another drought. The soil has been so depleted by the years of drought and excessive erosion that it will be centuries before it will become fertile again.

Angle of Repose: What Determines Slope Stability?

It is helpful for students to see the relationships among gravity, the stability of loose debris on a slope, and engineering considerations. In class demonstrations, I like to use simple materials, such as sand and gravel, to show these relationships. The instructor has two options here, one fairly simple and the other a bit more mathematically sophisticated. In the simplest approach, the instructor can simply define the angle of repose, as is done in the next paragraph, and illustrate the angle by placing a brick on a board and raising the board until the brick begins to slide. The instructor can also define the factor of safety, as is done in the second paragraph below, as a ratio that becomes less than 1 just when the brick begins to slide. In the more sophisticated presentation, the instructor can include sketches (on the board or as handouts) of the forces acting on a slope. This approach is explained in the last few paragraphs of this section.

To simplify the analysis of slope debris that might begin to slide downhill, consider a pile of sand or gravel (see Figure 6.3 in the Slide Set). No matter how much sand you put on the pile, it always assumes the same slope angle (about 30°–40°). All the cones of debris in the gravel quarry shown in the slide have about the same slope angles. If you are very careful, you can remove a bit from the base of the slope, but the slightest shaking—as during an earthquake—will cause the pile to reassume the same angle. (In fact, much mass movement does occur during earthquake-induced shaking.) If you were to measure the angle of the material once it has restabilized, it is likely that it would be about 30°–40°. This is the angle of repose of loose, unconsolidated sand at the Earth's surface. A state of repose means a state of rest, and the angle of repose is the maximum angle at which loose material comes to rest.

To understand why a given material has a certain angle of repose, we must consider both the driving forces (gravity) and the resisting forces (friction) acting on a slope. A brick can be used to demonstrate what happens to a particle on a slope. A brick placed on a horizontal board will not slide, but a brick placed on a vertical surface certainly will. Engineers use a measure called the factor of safety to determine whether or not a slope is stable. If the factor of safety is less than 1, the slope will be unstable; if it is greater than 1, the slope is likely to be stable. At some angle between horizontal and vertical, the brick will begin to slide. At a factor of safety of 1, the slope is on the threshold between stability and failure. The factor of safety is the ratio of the resisting forces (friction) acting to hold the brick in place and the driving forces (gravity) acting to move the brick downslope:

$$\text{Factor of safety (FS)} = \frac{\text{Resisting forces}}{\text{Driving forces}}$$

To illustrate the effects of friction, we will identify the forces acting on a brick that lies on a slope just less than that required for failure; for example, 30°. The only force exerted on the brick is gravity. That force can be expressed as the brick's weight, or the product of its mass and the acceleration of gravity (Figure 16.2). This force vector can be resolved into two components. One is acting perpendicular, or normal, to the slope and inhibits sliding. This is called the normal force (F_N). The other is acting parallel to the slope and is the driving force that causes sliding. It is called the shear force (F_S), because it tends to shear the particle along the failure plane.

The angle of the slope on which loose debris is resting determines the driving forces: the greater the angle, the greater the component of the force parallel to the slope. Once this angle becomes so large that frictional resistance cannot hold the debris in place, the factor of safety becomes less than 1, and failure occurs. For a brick, or a grain of sand in a pile, that angle happens to be about 35°.

From trigonometry, it can be shown that the normal force is proportional to the cosine of the slope angle, and the shear force is proportional to the sine of the slope angle. The slope angle is shown as the symbol q. On a horizontal surface, the slope angle is 0°; thus, the shear force is 0 and the normal force is at the maximum value—the cosine of 0 is equal to 1. On a vertical surface, the slope angle is 90°; thus, the shear force is at the maximum value—the sine of 90° is equal to 1. The normal force is equal to 0, because the cosine of 90° is equal to 0.

What holds the brick in place as the slope angle is increased is not just the normal component of the weight of the block, but also the frictional resistance along the base of the block. This resistance acts in a direction opposite to that of the shear force. A brick has more frictional resistance than a block of ice or polished metal, even if the normal force is the same. So the total resistance to motion is equal to the normal force times some frictional factor that is related to the material resting on the slope. In the equation for factor of safety, this product can be substituted for the numerator, the resisting force. Shear force can be substituted for the driving force:

$$F_S = \frac{\text{Resisting force}}{\text{Driving force}} = \frac{\text{Normal force} \times \text{Friction factor}}{\text{Shear force}}$$

At failure, the factor of safety is equal to 1, so this equation could be rewritten as follows:

$$\text{Shear force} = \text{Normal force} \times \text{Friction factor} \qquad \text{(Eq. 1)}$$

As the slope angle under the brick is increased, the shear force increases until eventually it reaches a threshold value and failure occurs. But if two bricks are stacked one upon the other, the normal force will be greater, and the shear force required for failure will also have to be greater. If three bricks are stacked together, the normal and shear forces at failure will be even greater. Thus, pairs of values for the normal and shear forces at failure can be plotted on an x-y graph, with the normal force on the x-axis to indicate that the shear force for failure is dependent on the normal force (Figure 16.3).

The data for this x-y graph plot on a straight line, for which the function is $y = mx$, where m is the slope of the line (for the case of a y-intercept of 0). Because y is equal to the shear force and x is equal to the normal force, the equation is

$$\text{Shear force} = \text{Normal force} \times \text{Slope of the line} \qquad \text{(Eq. 2)}$$

The slope of the line is equal to the shear force divided by the normal force, or the tangent of the slope angle, called ϕ.

$$\text{Shear force} = \text{Normal force} \times \tan \phi \qquad \text{(Eq. 3)}$$

But what exactly is this ϕ angle? From a comparison of Equations 1 and 2, it must be the same as the frictional resistance factor. At the angle of repose,

$$\text{Shear force} = \text{Normal force} \times \tan \phi \qquad \text{(Eq. 4)}$$

$$(\text{Weight}) \sin \theta = (\text{Weight}) \cos \theta \times \tan \phi \qquad \text{(Eq. 5)}$$

Thus,

$$\tan \phi = \frac{(\text{Weight}) \sin \theta}{(\text{Weight}) \cos \theta} \qquad \text{(Eq. 6)}$$

and the values of weight cancel each other, leaving

$$\tan \phi = \frac{\sin \theta}{\cos \theta} = \tan \theta \qquad \text{(Eq. 7)}$$

In other words, the frictional resistance has an angle equal to that of the slope angle at failure. The angle of repose is the angle at which maximum frictional resistance occurs, just before failure. For most loose, unconsolidated debris, that angle is between 30° and 40°.

Landscape Evolution and Darwin's Experiments with Earthworms

Charles Darwin, the British biologist famous for his ideas on evolution (see Chapter 3), devoted his last years to the study of earthworms and their effect on soils and the landscape. Using the phrase vegetable mould to refer to topsoil, Darwin titled his final book, one of his most charming, *The Formation of Vegetable Mould, Through the Action of Worms, With Observations on Their Habits* (1881). To alleviate the doubts of possible critics as early as possible, he wrote in the preface: "The subject may appear an insignificant one, but we shall see that it possesses some interest."

Darwin designed a series of clever and ingeniously simple experiments at his home to monitor the habits of worms and to determine how they contribute to denudation, or lowering, of the land surface over time. In the conclusion of his worm book, Darwin discussed the strength, ubiquity, and significance of worms:

> Worms have played a more important part in the history of the world than most persons would at first suppose. In almost all humid countries they are extraordinarily numerous, and for their size possess great muscular power. In many parts of England a weight of more than ten tons (10,516 kilogrammes) of dry earth annually passes through their bodies and is brought to the surface on each acre of land; so that the whole superficial bed of vegetable mould passes through their bodies in the course of every few years. . . . By these means fresh surfaces are continually exposed to the action of carbonic acid in the soil, and of the humus-acids which appear to be still more efficient in the decomposition of rocks. . . . Moreover, the particles of the softer rocks suffer some amount of mechanical trituration in the muscular gizzards of worms, in which small stones serve as mill-stones. . . . It is a marvellous reflection that the whole of the superficial mould . . . has passed, and will pass again, every few years through the bodies of worms. The plough is one of the most ancient and most valuable of man's inventions; but long before he existed the land was in fact regularly ploughed, and still continues to be thus ploughed by earth-worms.

By carefully observing worms in buckets in his office as well as in outdoor plots of ground, Darwin recognized that worms continuously churn soil through their intestinal canals, where they break particles into smaller bits and extract some nutrients before ejecting the rest as fine, soft castings that crumble apart easily. The spiral castings help give topsoil its loose, aerated, and powdery texture. This fine material is susceptible to erosion by wind and water on hillslopes.

Darwin then tested the possibility that many worms acting on a gently sloping (less than 10°) land surface could be major contributors to erosion. By repeatedly weighing worm castings as the worms moved downslope across a horizontal line 1 yard in length, he determined that in one year, 11.5 pounds of soil could be moved downhill across a line 100 yards in length. Darwin estimated that more than 50,000 worms must live under each acre of soil in regions such as Britain, and so their potential total impact is great. His computations showed that 0.2 inch of earth was brought to the surface by worms each year, and depending on the slope of the land, some fraction of this could be removed by erosion.

Darwin realized that most eroded debris ultimately makes its way to the ocean, which he called "the great receptacle for all matter denuded from the land." Comparing his field experiments to rates of denudation based on the amount of sediment carried into the ocean by the Mississippi River, Darwin marveled that the Mississippi's watershed is lowered 0.00263 inch per year, a rate that would reduce the entire drainage area to near sea level in only a few million years if the land were not also rising upward while being denuded. Furthermore, 0.00263 inch is a tiny fraction of the amount that surfaces each year in the castings of worms. With a keen sense of the power of geologic time, Darwin concluded that even if only a small fraction of the 0.2 inch of earth brought to the surface by worms each year "is carried away, a great result cannot fail to be produced within a period which no geologist considers extremely long."

CHAPTER 7

The Surface Water System

CHAPTER OBJECTIVES

Chapter 7 is the first of four chapters in Part III, Fluid Earth Systems. Chapter 7 focuses on water on continental surfaces, Chapter 8 on groundwater, Chapter 9 on water in the oceans, and Chapter 10 on gases in the atmosphere. Chapter 7 treats several parts of the surface water system: drainage basins, streams and rivers, and wetlands. The processes that cycle water through drainage basins and create streams are presented first, followed by a discussion of the hazards of flooding, and then a separate treatment of wetlands, wetland destruction, and wetland protection. The chapter concludes with an analysis of water quality and its protection. Specific objectives of this chapter are to

See how the surface water system has created an intricate network of streams and storage areas.

Assess the hazards of flooding, droughts, and pollution of surface water.

Examine the ways in which environmental changes have affected the surface water system.

Evaluate current efforts to protect and restore freshwater systems.

The global hydrologic cycle was introduced in Chapter 2, in a section on cycles in Earth systems. It is helpful if the instructor reminds students of the hydrologic cycle when beginning this section of the course. A more detailed treatment of the cycling of surface water on continents is presented in this chapter on page 198, in an example of how to use the hydrologic budget equation to calculate water availability for the city of Las Vegas in the arid American Southwest.

An example of global change is presented in Box 7.3, "Climate Change, Human Activities, and Surface Water Systems." Examples of regional change include discussion of the shrinking Aral Sea (page 194), the impact of damming the Colorado River (pages 192–193), the analysis of Colorado River flow from tree rings (Box 7.1), the shifting paths of the Mississippi River and its delta (pages 208–211), and the impact of channelization on streams (pages 212, 214–216).

We deliberately end this chapter on a hopeful note. Two views of a river in Massachusetts are laid out side by side. On the left, the river is shown as it was in the 1960s—highly polluted. On the right, we show the river now—substantially cleaned up as a result of years of legislation to protect clean water. We find that students can become quite disheartened during a course on environmental geology, and to prevent a feeling of hopelessness and doom we try to end lectures—as well as chapters—with positive information.

CHAPTER OUTLINE

1.   Surface Water Distribution

2.   Drainage Basins and Streams
      A.   Runoff Processes
      B.   The Role of Climate in Stream Discharge
      C.   Stream Channel Patterns and Processes
      D.   The Hazards of Flooding
      E.   Floodplains and Flood-Recurrence Intervals
      F.   Great Floods
      G.   Minimizing Flood Hazards

3.   Wetlands
      A.   Characteristics and Benefits of Wetlands
      B.   Coastal, Riverine, and Glacial Wetlands
      C.   Wetlands Destruction
      D.   Protecting and Restoring Wetlands

4.   Surface Water Resources and Protection
      A.   Freshwater Use
      B.   Surface Water Systems as Waste Disposal Sites
      C.   Regulations to Protect Drinking Water
      D.   Protecting Water Quality in Streams and Rivers

SUGGESTED LECTURE OUTLINE

The headings listed below are related to topics addressed in Chapter 7 and provide an alternative structure by which to present the material during classes and/or discussion periods. Items marked with one asterisk are treated in the following section on suggested lecture and discussion topics; those marked with two asterisks are treated in the section on demonstrations and case studies.

The Hydrologic Cycle Revisited
      Precipitation, evapotranspiration, and runoff
      The ability of rain to erode landscapes**
      Stormflow, baseflow, and stream discharge
            Stream gauging stations and the U.S. Geological Survey**
      Ephemeral and perennial streams
      How climate affects stream discharge

Drainage Basins and River Systems
       Drainage networks and drainage divides
       Stream channel patterns
       Stream restoration

Flooding
       Floodplains and floods
       Flooding on alluvial fans**
       Flood-recurrence intervals**
       Flood control structures
              Dams and their effects on streamflow and sediment load**
       The national flood insurance program

Changing River Systems
       The Mississippi River
       The Hwang Ho
       The Colorado River

Wetlands
       Importance to ecosystems and flood control
       Wetland types
       Wetland water budgets
       Wetlands destruction, protection, and restoration

Surface Water Quality and Protection
       Human consumption of fresh water
       Pollution of streams and rivers
              Point-source pollutants
              Non-point-source pollutants
       Federal regulations to protect drinking water

SUGGESTED LECTURE AND DISCUSSION TOPICS

Flood-Control Reservoirs and the 1993 Upper Mississippi River Basin Floods

The way the United States deals with flooding and flood control is one of the most important environmental and societal issues of the 20th century. Some experts predict that the National Flood Insurance Program might become the second greatest liability to the federal treasury, after the Social Security Program. Yet many people don't realize how this flood policy came about, what it is, what it can do, and what it can't do. Efforts to control flooding, including structural approaches such as channelization and flood-control reservoirs as well as nonstructural approaches such as floodplain zoning, are discussed on pages 209–216 of the textbook.

In the text that follows, we present the instructor with information about the performance of flood-control structures on the Mississippi River during the 1993 floods. The instructor can use this material to help students evaluate various approaches to flood control on a river system as large and powerful as the

Mississippi. Questions can be posed as to whether or not people should develop on floodplains, if wetlands should be protected for the purposes of storing floodwaters, or if towns that already exist on floodplains should be eligible for funds to relocate instead of to build levees and flood walls. Throughout this discussion, it is particularly helpful if the instructor explains hydrographs and illustrates how stream discharge as plotted in a hydrograph differs for preurban, urban, and post-flood-control reservoir conditions (as illustrated in Box 7.3). Our experience has indicated that students often have trouble understanding hydrographs unless we discuss them in class.

From mid-June through early August of 1993, widespread and severe flooding occurred throughout the upper Mississippi River basin, killing more than 50 people, destroying millions of acres of crops, damaging highways and roads, breaking many levees, severely eroding channel banks and hillslopes, and depositing sediment over large areas (see Figure 7.18 in the textbook). As a result of a six-month period of greater than normal precipitation, soils throughout the region were already saturated when a sequence of intense rainstorms began in late June. Little or no infiltration occurred, and most of the precipitation became direct storm-water runoff, rapidly raising the water levels of numerous streams that were already at bankfull conditions. Flood-control reservoirs were at or near capacity (see Figure 7.19a in the textbook). One storm after another pelted the area for nearly two months, and one tributary after another swelled and disgorged its load into the main Mississippi River trunk stream. Peak discharges exceeded 100-year flood levels at 46 gauging stations and were greater than any previously recorded values at 42 stations.

As this period of flooding illustrated, it is the number of tributaries and the vast areas they cover that make the Mississippi so threatening when it floods. The Mississippi watershed drains more than 100,000 tributaries over about 1.25 million square miles of land, from the Rockies in the west to the Appalachians in the east. Its largest tributary, the Ohio River, drains the Appalachian Mountains and lies within the track of cyclones that often move northeast from the Gulf of Mexico. The Mississippi trunk stream itself joins the Missouri and Illinois rivers just upstream of St. Louis, Missouri. All three of these rivers drain the upper midwestern United States, where an eastward-flowing jet stream drawing moist, warm air from the Gulf of Mexico converged with cool air masses from Canada throughout the spring and summer of 1993. The result was persistent, unstable air masses and rainfall. Fortunately, the Ohio River basin was not in the path of these storms, and the Ohio River was flowing below normal levels at that time, or else flooding would have been far worse downstream of its confluence with the Mississippi.

Flooding would also have been far worse if not for the storage of large volumes of water in dozens of flood-control reservoirs, most of which are located within the Missouri River basin. A flood-control reservoir is designed to store part of the stormflow during a flood and to release it later in order to reduce the flood peak downstream. A stream with one or more such reservoirs is referred to as a controlled stream; one without storm-water reservoirs is called an uncontrolled stream.

A convenient measure of storm response is the hydrograph, a plot of the rate of water flow in the channel over a time interval beginning with the storm event and continuing until response to the storm has diminished (see Box 7.3 in the textbook for an example hydrograph). A hydrograph can be used to illustrate the differences between controlled and uncontrolled streams during flooding, as shown in Figure 7.20 in the textbook. In an uncontrolled stream, each tributary adds more water to the trunk stream, resulting in a progressive downstream swelling of both peak discharge and total discharge. In contrast, reservoirs along a controlled stream can hold water until the danger of flooding on the main stream has passed or can minimize the extent of downstream flooding.

Flood-control reservoirs in the Missouri River basin reduced the discharge at St. Louis by more than 211,000 cfs in July 1993. Without the reservoirs, many more levees would have been breached, and more cities would have been flooded. Nevertheless, the amount of death and destruction that occurred despite the levees and reservoirs suggests that the only long-term solution to flood hazards is to dissuade development in flood-prone areas. Some communities, such as Patterson, Kansas, have taken this tack by using federal relief funds to rebuild their towns at higher elevations after the 1993 floods.

DEMONSTRATIONS AND CASE STUDIES

Please see the previous section, Suggested Lecture and Discussion Topics, for an additional case study on flood-control reservoirs and the 1993 upper Mississippi River
basin floods.

How to Measure Raindrops and Calculate Rainfall Intensity

Rainfall intensity is a function of raindrop size. In most temperate climates, rainfall intensities are much less than 75 mm/hr; in tropical countries, they can be as high as 150 mm/hr. The maximum rate measured was in Africa, at 340 mm/hr, but this rate was sustained for only a few minutes. By comparison, typical infiltration capacities of a loose, silty soil are 25 mm/hr and those of a sticky clay are less than 4 mm/hr.

A simple experiment was devised a number of years ago to sample raindrops and determine the approximate rainfall intensity during a storm. Fill a cake pan with fresh, sifted white flour (no lumps), put a cover on the pan, and take the pan outside in a rainstorm. Aluminum foil works as a cover, but it shouldn't touch the flour. Once you are standing in the rain, check your watch, remove the cover, and wait several minutes. Any time period from 2 to 6 minutes should work. The important point is to wait long enough for many raindrops to cover the surface of the flour, but not so long that the surface becomes soaked and matted down. Replace the cover and check your watch to note the length of your sample time. Set the cake pan aside for a day or two to allow the drops to evaporate. Keep the cover on the pan to protect the surface.

While the drops sit in the flour, and before they evaporate, they absorb a thin coating of flour because of surface tension. As the water inside the flour coating evaporates, the coating thickens and a small, hard ball remains. You can sift the flour carefully to remove the raindrop balls, then assess their size distribution. The greater the intensity of the rainstorm, the larger the balls. In general, the largest size of a raindrop is about 9–10 mm for a very intense rainstorm. You can calculate the total rainfall volume for an area the size of your cake pan for the time period you monitored the rainfall. The total volume of rain is merely the sum of the volumes of all raindrop balls. You can calculate the volume using the equation for the volume of a sphere ($V = 4/3\pi r^3$, where r is the radius of a raindrop, or one-half its diameter). Divide this volume by the area of the pan to get the thickness of the rainfall event (in millimeters) for the sample time period, thus giving rainfall intensity in mm/min or mm/hr per unit area. Compare this value to that reported in your local newspaper's weather report.

Example Calculation:

Sample time = 1 minute

Cake pan area = 9 inches by 9 inches (229 mm × 229 mm) = 52,441mm^2

20 raindrops each have a diameter of 5 mm, so

Volume of 20 raindrops = $(20)(4/3\pi)(2.5^3)$ mm^3 = 1309 mm^3

30 raindrops each have a diameter of 8 mm, so

Volume of 30 raindrops = $(30)(4/3\pi)(4^3)$ mm^3 = 8042 mm^3

15 raindrops each have a diameter of 10 mm, so

Volume of 15 raindrops = $(15)(4/3\pi)(5^3)$ = 7854 mm^3

Total volume of raindrops = sum of all raindrop volumes = 1309 + 8042 + 7854 = 17,205 mm^3

Intensity of rainfall in 1 minute = volume/area = 17,205 mm^3/52,441 mm^2 =
0.328 mm/min, or 0.013 inch/min
which is equal to 19.68 mm/hr, or 0.748 inch/hr

Getting Water Resources Data from the United States Geological Survey

The nation's principal water-science agency is the United States Geological Survey (USGS), an agency within the Department of the Interior that employs nearly a thousand geoscientists. USGS operates an extensive hydrologic data network that includes 7300 continuous-record streamflow-gauging stations. Some streamflow records extend back to the 19th century, and this database is one of the world's most continuous and detailed records of hydrologic data. Because the USGS is federal and its mission is to serve the nation, hydrologic data are available to anyone, at little or no cost. The main source of information on river discharges is an annual government publication titled *Water Resources Data for (State): Part I, Surface Water Records,* which is available for each state. Recently, the USGS has made all this data available at their web site (http://h2o.usgs.gov/).

Before 1960, data were published in an annual series of Water Supply Papers for 14 water regions, with a document for each region titled *Surface Water Supply of the United States.* Surface water records in these documents contain detailed information for each streamflow-gauging station, including mean daily discharges in cubic feet per second (cfs), maximum yearly flows, gauge height and peak flow data necessary to construct rating and flood-frequency curves, and notes on the location of the gauging station and its drainage area. These documents are available in most libraries in the government document sections or from USGS offices. Today, historic and real-time water resources data are available in electronic format, on floppy disks and CD-ROMs.

As a classroom exercise, it is useful to show students how to locate the real-time and historic data for a gauging station on a local river. The instructor can go to the USGS Water Resources home page (http://h2o.usgs.gov/) and follow the instructions to locate the station of interest by using a map or list. Once at the station, the instructor can show real-time stage and discharge data. If large floods occur somewhere in the United States during the semester, the instructor can go to this web site and show the hydrograph of rising floodwaters. In addition, the USGS commonly posts photographs and relevant details of current floods along with associated hydrographs during major floods.

The Los Angeles Basin and Flooding on Alluvial Fans

Alluvial fans, like deltas, are cone-shaped landforms created by deposition at the downstream ends of channels where they emerge onto plains of low gradient. Unlike deltas, they are not restricted to the shores of coastlines or lakes, and they form above rather than below standing water. The streams that form alluvial fans do not remain fixed in the same spot for long, so it is very difficult to predict flood hazards for such streams. They are as likely to gouge out the sediment in their way and branch into multiple channels as to overflow their banks (Figure 17.1). Such streams are said to have locational instabilities, and they pose a unique set of problems for the engineers, hydrologists, and geoscientists who map special flood hazard areas.

Two other aspects of alluvial fans make them especially hazardous to people who live on them. First, alluvial fans are most often found along mountain fronts. Consequently, the sediment excavated from the mountains to build the fans is coarse—as coarse as boulders the size of cars in some cases. The streams that carry such debris are usually steep and prone to debris flows and mudflows, which can be much more destructive than overbank flooding. A second problem is that most actively forming alluvial fans occur west of the Rocky Mountains, along the fronts of such tectonically active mountains as the Wasatch Range bordering Salt Lake City and the San Gabriel Mountains bordering the north side of the Los Angeles basin. The streams that form alluvial fans in these locations pose problems for flood hazard specialists, because most of the standards and guidelines established by FEMA (the Federal Emergency Management Agency) for determining flood-prone areas were based on streams that do not have locational instabilities. For example, the Susquehanna River at Harrisburg, Pennsylvania, like many other eastern rivers, overflows its banks during flooding. The Los Angeles River, on the other hand, is more likely to enlarge its channel by cutting its banks as it flows across its alluvial fan surface. Of course, the millions of inhabitants of Los Angeles prefer a river that stays put, so the Los Angeles Metropolitan Water District has devised an intricate system of debris basins and channelized streams designed to prevent unwanted channel wandering.

At present, FEMA and the U.S. Army Corps of Engineers are struggling to deal with such problems. Rapid growth in sunbelt states such as Arizona, California, and Nevada, which are speckled with alluvial fans, only worsens the problems, for FEMA must develop new guidelines and regulations while the areas they seek to understand are in the midst of rapid development.

Dams on the Colorado River and their Effect on Water and Sediment Flow

Modern dams and aqueducts have added a new dimension to the possibilities for change in the system dynamics of a river. A classic example is found on the Colorado River in the semiarid southwestern United States. Reservoirs behind numerous dams create large surface areas where water is lost by evaporation. For example, in the Powell Reservoir behind Glen Canyon Dam, 9 percent of the river's water is evaporated and the remaining water becomes saltier. After the dam was closed in 1962 and brackish waters were returned to the river from irrigated fields, water salinity at the river's mouth increased from 800 mg/liter to 1500 mg/liter. Reservoirs also trap sediment and eventually fill completely with debris, greatly decreasing the sediment load of the dammed river. Canals and aqueducts are used to remove large amounts of surface water from a drainage basin and transport it to another watershed for human use.

In the cases of damming and water transfers, input from precipitation and sediment have not necessarily been altered, but the flow of water and sediment along the channel and the output at the mouth have decreased markedly. The discharge of the Colorado River has declined continuously since the late 1930s as more and more of its water is diverted for irrigation and municipal drinking water supplies. Its sediment load decreased abruptly when the Hoover Dam was completed in the mid-1930s, and again in the 1960s as other dams were completed (Figure I7.2).

The side effects of diminished in-stream flow and damming on environmental systems make it clear that ecosystems and habitat suffer from decreased water and sediment loads. As water flow and sediment loads decrease, a stream adjusts to the changed conditions. First, at the upstream end of a reservoir, the stream encounters a reduced gradient and water velocity because water has ponded behind the dam. Because the river's capacity to carry sediment has been decreased, it absorbs this change by depositing its load. This process is called aggradation. Downstream of the dam, water depleted of its sediment load is released into the channel at its original gradient. As a consequence, the flow has more capacity than needed to transport sediment, and it is able to scour debris from point bars, the floodplain, and the channel bed. In other words, the river channel degrades. The debris eroded from the river channel is generally fine-grained because large discharges rarely occur after dams and flood-control works are built. As a result, rivers such as the Colorado are unable to flush out coarse deposits of gravel and boulders that were moved during past floods. These types of change have been documented on many rivers throughout the world, including the Colorado, Brazos (Texas), Nile, and Missouri rivers.

The diminished flow also has political consequences. The Colorado River is often merely a salty dribble at its outlet into the Gulf of California in Mexico. In the 1960s, the Mexican government protested that its allocation of water from the Colorado (guaranteed by the Mexican Treaty of 1945) was too brackish to be fit for human consumption or irrigation, so the U.S. government agreed to build and maintain a desalination plant in Yuma, Arizona. This desalination plant is the largest in the world.

A number of legislative acts and treaties between southwestern states and Mexico brought the Colorado River to this plight. In 1928, Congress passed the Boulder Canyon Project Act, giving the Bureau of Reclamation and Army Corps of Engineers authorization to construct the giant Hoover Dam, numerous small dams, and several canals. Hoover Dam is the highest in the nation (700 feet) and was cited by the American Society of Civil Engineers as one of seven civil engineering wonders in the nation. Electricity generated by dams along the Colorado River is sold throughout the southwest and used to subsidize the supply of irrigation water to farmers in the surrounding deserts of California, Arizona, Utah, Colorado, and Mexico. The Boulder Canyon Project Act guaranteed a supply of 4.4 million acre-feet of water per year to California, and the Colorado River aqueduct was built to transport water to southern California. Without this allocation, the growth of cities such as Los Angeles would have been severely curtailed. Today, water from the Colorado supplies more than 12 million people in California.

The state of Arizona protested the building of Hoover Dam, since its own allocation of Colorado River water is only 2.8 million acre-feet. Furthermore, California was guaranteed to receive its 4.4 million acre-feet even during drought years. Arizona once relied largely on groundwater resources to supplement its water needs, but these resources have been depleted rapidly by its booming population growth since World War II. As a result, the states of Arizona and California, as well as other states and native American tribal nations that use water from the Colorado, have been mired in a series of complex court cases since the 1950s as they battle over the rights to each precious acre-foot of water.

Using Statistics to Calculate Flood Recurrence Intervals

The concept of the 100-year flood is fundamental to flood zoning and insurance in the United States. For this reason, it is worth the effort to explain to students how the discharge of the 100-year flood is determined. The procedure is simpler than many people imagine, and it can be grasped by the students in an introductory course if it is presented sequentially. In our classes, we find it useful to work with historic data for a nearby stream gauging station from the USGS Water Resources web site. In your own class, you could choose data from any nearby station or work with the data for a stream that might be in flood at the time of your lecture on flooding. The following text can help you to present a discussion of the 100-year flood. When it comes time to plot the data, we simply do it before class and make a transparency or call up the relevant file on a computer projector. Although the text below notes that special paper is used to make flood-frequency plots, for the purposes of this discussion you can use ordinary semi-log graph paper (or computer graphing) and explain to students that a special type of logarithmic paper normally is used by engineers and hydrologists.

We cannot look at a stream and predict how great its flow will be during the next year. But we can examine the frequency of past flows and estimate the probability of a given discharge being equaled or exceeded in any one year, based on its past record. This probability is known as the exceedance probability, and an obvious assumption of this method is stationarity of the mean. In other words, a statistical measure of past events, such as the mean value, should not vary significantly over the period for which predictions are made. An example of a phenomenon that might make such an assumption invalid is a climatic shift due to increasing occurrence of El Niño events. Nevertheless, because this is the only method available for meaningful flood prediction, it is accepted for use in the United States.

The flood-frequency analysis procedure authorized by FEMA for use in floodplain zoning and insurance can be divided into the five steps presented below. With data from the USGS Water Supply Papers or web site (http://h2o.usgs.gov/), you could do a similar analysis for this or any other river in the United States for which flow data are available.

Step 1: Get a sample of discharge events from a stream gauging station.

The pattern of discharges documented by a sampling of all events is used to estimate what might happen in the total population of all conceivable events at that location, thus enabling predictions of future events. A common sampling used in flood-flow analysis is the set of annual maximum discharges. The annual maximum discharge is the largest event that occurred each year, and the total sample size is equal to the number of years of flood-flow record for that station. In the United States, few stations have records going back more than 90 years. We will examine the Nagle Street station on the Susquehanna River near Harrisburg, Pennsylvania, as an example. Continuous records for this station have been kept since 1890, so the sample size in 1993 was 104.

Step 2: Rank the events by magnitude, with the largest event assigned the number 1 and the smallest event a number equal to the number of years of record.

At Nagle Street, the greatest annual maximum discharge occurred in 1972, during Hurricane Agnes, and was equal to 1,0163,00 cfs.

Step 3: Calculate the probability of exceedance in any given year for each event.

The probably of exceedance (PE) is calculated using the following equation:

$$PE = \frac{m}{(n+1)}$$

where PE is the probability of exceedance; m is the rank, or magnitude, of the event, from largest to smallest; and n is total number of samples, or years of record.

Thus, for the largest flow on record at Nagle Street:

$$PE = \frac{m}{(n+1)} = \frac{1}{(104+1)} = 0.0095, \text{ or } 0.95 \text{ percent}$$

The larger the event, the smaller its probability of exceedance in any given year.

Step 4: Calculate the recurrence interval for a given event.

The recurrence interval, RI, is the reciprocal of the probability of exceedance:

$$RI = \frac{1}{PE} = \frac{(n+1)}{m}$$

Thus, for the Hurricane Agnes flood at Nagle Street:

$$RI = \frac{1}{0.095} = 105 \text{ years}$$

The larger the event, the greater its recurrence interval.

Step 5: Graph the data and use the graph to estimate probabilities of exceedance and recurrence intervals from a flood-frequency curve.

A limitation of the flood-frequency method is the sample size. In the Nagle Street example, the largest event to occur in the 104 years of recorded history is not necessarily the flood that has a recurrence interval of 104 years; it just happens to be the largest flood recorded in that time period. With more years of data, the event might be estimated as a 150-year event or even a 500-year event.

To address this problem, the probability of exceedance data are plotted on a special type of statistical graph paper, known as log-Pearson type III, as recommended by FEMA. Flood discharge is plotted on the vertical axis and probability of exceedance on the horizontal axis (see Figure [a] in Box 7.2 of the textbook). The reciprocal of probability of exceedance, the recurrence interval, can be plotted on the opposite horizontal axis, or a double x-axis plot can be used to show both parameters. A best-fit curve is computed for the data, and this curve can then be used to estimate either the discharge for any given probability of exceedance or the recurrence interval.

For example, from the Harrisburg flood-frequency data, the large discharge for Hurricane Agnes is estimated to have a recurrence interval of 200 years when using a flood-frequency plot, rather than the 105 years estimated from the equation used in Step 4.

CHAPTER
8

The Groundwater System

CHAPTER OBJECTIVES

Chapter 8 is similar to Chapters 6 and 7 in its organization. It begins with fundamental information about the occurrence of water underground, the flow of water from one place to another as a result of an energy potential, porosity and permeability of aquifers, and groundwater chemistry. The value of groundwater as a resource follows this section and in turn is followed by a section on groundwater hazards, including sinkholes, land subsidence from lowered water levels, and intrusion of salt water along coastlines. As in previous chapters, this one closes with discussion of groundwater pollution and its cleanup. Types of pollutants and their migration are described, and the few environmental laws that exist to protect or clean up groundwater are presented. The chapter objectives are to

Find out how groundwater gets underground and flows from one place to another.

Consider how groundwater is used and the environmental changes that have affected its quality and quantity in recent years.

Investigate hazards associated with the extraction of groundwater, such as sinkholes, subsidence, and contamination by seawater.

Examine how groundwater can be cleaned if it becomes contaminated.

This chapter has even more applied geology than some of the others in this textbook. We present numerous case studies, including contaminant plumes at Cape Cod (pages 234 and 254–255) and groundwater resource management in Los Angeles (pages 246–247). This chapter, like Chapter 7, ends on a hopeful note, in this case by illustrating advances in bioremediation for cleaning up groundwater contaminated with petroleum products.

CHAPTER OUTLINE

1.   Groundwater in the Hydrologic Cycle
        A.   The Water Table
        B.   Effect of Elevation and Pressure on Groundwater Movement
        C.   Porosity and Groundwater Storage
        D.   Permeability and Groundwater Flow
        E.   Aquifers
        F.   Groundwater Chemistry

2.   Groundwater as a Resource
        A.   Gaining Access to Groundwater
        B.   Groundwater Resource Management

3.   Groundwater Hazards
        A.   Solution Caverns and Sinkholes
        B.   Land Subsidence From Groundwater Mining
        C.   Intrusion of Salt Water

4.   Groundwater Pollution and Its Cleanup
        A.   Types and Sources of Groundwater Pollution
        B.   Migration of Groundwater Pollution
        C.   U.S. Laws Governing the Quality of Water Resources
        D.   Groundwater and Aquifer Restoration

SUGGESTED LECETURE OUTLINE

The headings listed below are related to topics addressed in Chapter 8 and provide an alternative structure by which to present the material during classes and/or discussion periods. Items marked with one asterisk are treated in the following section on suggested lecture and discussion topics; those marked with two asterisks are treated in the section on demonstrations and case studies.

Aquifers and Groundwater Flow
        Porosity and permeability**
        The water table and Darcy's law
        Groundwater budgets: recharge and discharge**
        Confined and unconfined aquifers
        Artesian flow
        Groundwater regions of the United States
        Springs and geysers

Wells and Groundwater Pumping
        Groundwater flow to a well
        Hand-dug wells, qanats, and modern irrigation systems
        Cones of depression and drawdown

The Groundwater Resource
    Groundwater resource management
    Aqueducts, injection wells, and spreading basins
    Groundwater mining and land subsidence

Groundwater Quality
    Groundwater chemistry
    Intrusion of salt water along coastal areas
    Groundwater pollution**
        Miscible contaminants (e.g., road salt)
        Immiscible contaminants (e.g., petroleum products)
    Contaminant plumes
    Leachate

Groundwater Laws, Protection, and Cleanup
    Superfund
    Aquifer restoration
    Bioremediation

SUGGESTED LECTURE AND DISCUSSION TOPICS

Nonrenewable Groundwater: Mining Saudi Arabia's Fossil Water

Groundwater is an increasingly valuable resource. As described in the text (page 245), it accounts for nearly 40 percent of all water used in the United States, and the amount has been increasing steadily for decades. Some cities rely completely on groundwater. For this and other reasons, efforts to protect and clean up groundwater have increased also.

Yet in some places, groundwater is used at such rapid rates that the aquifers containing it are exhausted of their supply. In essence, the water is mined from the aquifer. Groundwater mining is mentioned briefly on page 246 in the textbook, and Figure 8.13 illustrates 7 meters of a well that has been exposed due to subsidence in Mexico City, where groundwater mining is a serious problem. Otherwise, the chapter has little else on groundwater mining. The third thought question at the end of the chapter asks students whether groundwater is a renewable or nonrenewable resource. The question is important and is worth discussing with the students. To assist in such a discussion, we provide the following information about groundwater mining in Saudi Arabia. This example is particularly interesting because of its links to climate change and oil wealth.

In some areas of the world, recharge rates are now so low that the supply of groundwater is considered to be a fixed stock resulting from past recharge during wetter climatic periods. Groundwater in an area of little or no current recharge is called fossil water. For example, radiometric dating indicates that the groundwater in Saudi Arabia is 15,000 to 40,000 years old and that it accumulated under the much cooler conditions that existed throughout the world during the last glacial period (about 15,000 to 50,000 years ago). Surface water was more abundant in what is now Saudi Arabia at that time, and much groundwater recharge occurred. In areas of fossil groundwater, aquifer discharge from pumping results in a net loss of the groundwater stock, unless efforts are made to artificially recharge the aquifer. If pumping rates exceed rates of natural and artificial replenishment, groundwater overdraft occurs. This overdraft is called groundwater mining.

As a result of its geologic and environmental history, Saudi Arabia is underlain by immense reserves of both oil and groundwater. These reserves are nonrenewable, however. The oil is nonrenewable simply because it forms at rates that are slow relative to the scale of a human lifetime. Most Saudi groundwater, nearly 90 percent, is nonrenewable because it accumulated during times of much cooler and wetter glacial climatic conditions. Today, Saudi Arabia receives so little rainfall that very little aquifer recharge occurs.

This oil-rich nation has made an immense effort to diversity its economy, using its oil profits to subsidize the development of its groundwater resources in order to boost agricultural productivity and national self-sufficiency. In 1975, only 150,000 hectares of land were farmed; by 1988, this had risen to about 3 million hectares. Wheat production increased from 3000 metric tons in 1976 to 3 million metric tons in 1989, more than two-thirds of which was exported to other countries. Government subsidies to grow this wheat cost 8 times as much as it would cost for the country to import it!

The main source of Saudi Arabia's wheat has been the groundwater used to irrigate fields. In 1988, Saudi Arabia used 20,520 million cubic meters of water. The stock of groundwater diminished from 497,500 million cubic meters in 1980 to 385,000 million cubic meters in 1989—a reservoir depletion of 23 percent! Some experts estimate that the nation's fossil groundwater will be completely exhausted by the year 2007. In an effort to avoid this disaster, engineers are developing sophisticated irrigation techniques and building dozens of desalination plants that remove salts from brackish and marine waters. Although Saudi Arabia has the largest desalination program in the world, it is costly: $2.70 per cubic meter of water produced. As a consequence, this method is not yet economically feasible for growing wheat or other crops that demand moderate or large amounts of water. Nevertheless, Saudi Arabia has immense reserves of one resource that provides them with huge revenues—oil. It is possible for this nation to do things that others in arid and semiarid regions cannot, simply because of the amount of money it can afford to spend for water development.

DEMONSTRATIONS AND CASE STUDIES

Please see the previous section, Suggested Lecture and Discussion Topics, for an additional case study on mining Saudi Arabia's fossil groundwater.

Water Pollution: Misplaced Matter

This demonstration provides a useful way to get students to think about water contamination and such drinking water standards as the EPA (Environmental Protection Agency) maximum contaminant levels. The demonstration takes about 5 to 10 minutes and can be followed by a discussion of maximum contaminant levels for various pollutants.

To begin, fill one large, clear glass with rainwater. Pose the following question to your students: "Is this water pure?" Someone might request that you explain the meaning of pure. Explain that pure water contains no substances other than hydrogen and oxygen. Distilled water is pure water. Although students might guess that rainwater is likely to be pure, you can explain that it probably is not. As rainwater falls from thousands of feet above the Earth, it passes through clouds and atmospheric gases containing industrial pollutants and exhaust fumes. Because it is a solvent, water dissolves nitrogen, carbon dioxide, and many other gases. You can use a hot plate to heat the glass of rainwater and demonstrate how tiny bubbles rise to the surface as these gases—which are less soluble at higher temperatures—escape again to the atmosphere.

If the rainwater fell through the atmosphere above a large urban and industrial area, it would probably contain far more impurities than if it fell to Earth above the Arctic Circle. Surprisingly, however, even rainwater in the Yukon Territory has been found to contain an insecticide (toxaphene), apparently blown in from the former Soviet Union.

Next, fill four more glasses with tap water. Label the glasses 1 through 5, starting with the glass containing rainwater. The second glass will contain just tap water. To glass number 3, add a quarter teaspoon of salt. To glass number 4, add a tablespoon of salt. To glass number 5, add one drop of gasoline. Now label a sixth glass and fill it with bottled water purchased from a store. Pose the following question to the class: "Which of these glasses of water is likely to be the safest to drink?"

Ask the students if they know the source of the tap water (the public water supply). The tap water is likely to be groundwater, which supplies the majority of water to urban areas, or water from a nearby stream or river. If its source is a stream or river, it has probably passed through numerous settlement, coagulation, and filtration basins and then been disinfected at a municipal water treatment plant. Explain to students that if they were to look at the water in most rivers and streams near urban areas, they probably would not want to drink it before it underwent all of those procedures. During field trips, we often point out the water in streams passing through urban areas and stop to examine the intake of water at our local water treatment plant. Students can see the accumulation of tires, plastic, and other garbage that is trapped by the first of a series of filters. If you teach near a water treatment plant, it is a good idea to point out the facility during a class trip or show students its whereabouts on a map.

Ask the students to consider which glasses of water they would feel safe drinking and which they would be afraid to drink. Pass around the glasses and encourage the students to smell them—which one has an offensive odor? They probably would feel safe drinking the rainwater, bottled water, or tap water. All will say with certainty that they would not drink a glass of water containing even a drop of gasoline, which can be smelled even in such a tiny amount. They might consider drinking the water containing a quarter teaspoon of salt, but they would not want to drink the water containing a tablespoon of it. Salt is vital to life, but too much of it can cause illness or even death. On the other hand, the students might readily add a tablespoon of sugar to a glass of water, such as when making a glass of iced tea. Too much sugar might make the drinker feel ill, although it would not be as harmful as too much salt.

Some impurities are pollutants in very small amounts; others can be ingested in much larger amounts before becoming hazardous. For these reasons, chemists call substances such as the contaminants in drinking water "misplaced matter." Gasoline in a car is not unwanted, but that in drinking water is. Chemists also like to remind us that there is a little bit of everything in anything. In other words, few substances are pure. In terms of human health, however, we must decide how much is too much and how to determine that amount.

As a result of the Safe Drinking Water Act, passed in 1974 and amended in 1986, the EPA has established drinking water standards for all public water supplies in the United States (see Table 7.5 in the textbook). Maximum contaminant levels (MCLs) and maximum contaminant level goals (MCLGs) have been identified for several dozen substances. An MCL is an enforceable standard that sets the concentration at which a given substance is considered safe to human health if a person is exposed to that level of the substance over a lifetime. For example, the MCL for lead, which can cause neurological disorders, is 0.05 mg/liter; for trichloroethene, which is a possible carcinogen, the MCL is 0.005 mg/liter.

An MCLG is a nonenforceable standard that is even more stringent than an MCL but is considered unlikely to be attainable because of technological or economic limitations. For example, ideally, a possible carcinogen such as trichloroethene would not be present in any measurable amount in our drinking water. Its MCLG, therefore, is 0 mg/liter. But because trichloroethene is found in many water supplies and because it would be very costly to remove all traces of it, the MCL is set at a level that is economically feasible to achieve.

Getting Water Out of Sand

Gravel and glass beads can be used to demonstrate a variety of concepts related to groundwater storage, aquifer permeability, and contaminant transport. The following demonstrations are simple, rely on readily available and inexpensive materials, and require little preparation time. In our experience, they have been very effective pedagogical tools during lecture periods, and students often remark how valuable it was to see the demonstrations rather than just to hear a lecture on the same topics. We do the demonstrations with a substantial amount of student interaction and feedback, and we pose questions throughout the lesson.

Materials

Four 1000-ml (1-liter) beakers
Glass beads (6 mm in diameter, 600 ml in volume) and/or similar coarse gravel
Glass beads (3 mm in diameter, 600 ml in volume) and/or similar fine gravel
One 100-ml burette
Water and red vegetable dye
Two pieces of tygon tubing (diameter smaller than sediment; e.g., 2 mm; about 2–4 feet in length)

Brief Introduction to Terms and Concepts

We begin our classes on groundwater with a brief introduction to porosity, hydraulic conductivity, and aquifers, and we define each of those terms. We comment that it is surprising how much water can be stored underground in the tiny pore spaces between sediment grains or along cracks in rocks. If all the pore spaces that are filled with water in the subsurface of the United States were connected, they would form one large cavern 57 meters (188 feet) high underlying the entire country.

Any water-holding and water-bearing porous reservoir is an aquifer. For an underground aquifer, two characteristics—porosity and hydraulic conductivity—determine the amount of water it can hold and yield. Porosity is the ratio (usually given as a percentage) of void space to solid rock in a segment of earth.

The greater the porosity of a geologic material, the more water it can hold. Hydraulic conductivity is a measure of the rate at which a porous medium can transmit water and is given in units of distance over time (for example, meters per day). The greater the hydraulic conductivity of a geologic material, the more water it can yield to a pumping well or spring. Hydraulic conductivity is related to the fluid moving through the medium as well as to the nature of the porous material. For example, an oil that is more viscous than water would be transmitted more slowly. A weakly cemented sandstone would transmit water more rapidly than a sandstone with cement filling most pore spaces. The weakly cemented sandstone is more permeable than the well-cemented sandstone. So, permeability is related to hydraulic conductivity but is not synonymous with it. It is worthwhile to make this distinction to students and to point out that water is not the only fluid moving through the subsurface.

Demonstrating Porosity

Once these basic concepts have been defined, the instructor holds up a 1000-ml beaker filled to the 600-ml level with either 6-mm glass beads or coarse, rounded gravel (6 to 10 mm in diameter). (We use beads that we buy from a nearby glassware supply company.) The instructor then holds up a second beaker filled to the same level with 3-mm beads or fine gravel (3 mm in diameter). We ask the students, "Which 'aquifer' can store the most water—the one with the coarser or finer sediment?"

We write a tally of student answers on the board, with a column for those who say the coarser sediment, one for those who say the finer material, and one for those who say that both will be able to store the same amount of water. During five semesters of using this demonstration, we have found that nearly every student thinks the coarser beads or gravel will have greater porosity than the finer beads or gravel.

At this point, we ask the students for advice on how one might determine porosity. Someone will usually suggest that one way to do so is to pour water (dyed pink with red food coloring to make it easier to see) into the beaker and measure how much water is added. We use a 100-ml graduated cylinder so that we can measure volume to the nearest milliliter. We call a student to the front of the room to do the pouring and measuring. If we use all 100 ml, we refill the graduated cylinder and keep pouring. Usually, we can add about 150 to 200 ml of pink water to each beaker before it reaches the top of the sediment (or beads) at the 600-ml mark. If 150 ml of fluid can be added, that means that the sediment has a volume of pore space that is 25 percent of the total volume (150 ml/600 ml = 0.25). It is helpful to write this calculation on the board and to explain its relationship to the definition of porosity.

The demonstration illustrates that about the same amount of water can be added to both beakers of sediment. Slight differences of about 10 to 20 ml might occur, but these are accounted for by a number of sources of error in the experiment. We tell the students that porosity would be about 26 percent even if one packed beach balls together into a giant beaker. All closely packed sediments that are well sorted and rounded, regardless of their size, have a porosity of about 26 percent. We make a sketch on the board (or use an overhead transparency) to illustrate close packing (also called rhombohedral packing) with one layer of spheres tucked into the pockets formed by the next layer. If the packing is looser, with spheres stacked directly atop one another, the porosity will be greater, but the maximum porosity possible is about 47 percent. Typical values of porosity for sediments and rocks are presented in the textbook in Table 8.1. It is helpful to refer to this table at this point in the demonstration.

Demonstrating Hydraulic Conductivity

We ask students to think about the nature of the pore spaces in the sediments. Are they connected to one another? How interconnected are they? Would these sediments be permeable and yield water readily? Next, we ask students if they can suggest a way to get the water out of the sediment. In our experience, one or more students will suggest siphoning the water just as one would siphon gas from a tank. At this point, we pull out tygon tubing and push it deep into the beaker so that its end touches the bottom of the beaker. We put the other end of the tubing into an empty beaker. The goal of this part of the demonstration is to get as much water out of the sediment as possible. We have someone hold the beaker of coarser sediment about two or three feet above an empty beaker. We start a flow of water into the tubing by creating suction on the end in the empty beaker. The pink fluid will begin to flow down the tube and fill the beaker. Once flow has ceased, we drain the tubing and pour the fluid from the beaker into the burette to measure its volume. Usually, we get about 100 to 140 ml of water, even though we put more than that into the sediment. The reason for the difference is that some of the water has been retained by the sediment as a result of surface tension between the grains and the water molecules. It is not possible to extract all the water from an underground reservoir, just as it is not possible to get all the oil out of the ground by pumping a well.

Now we ask the students whether we will get more water, less water, or the same amount of water from the finer sediment. Because of their previous experience with guessing the porosity of the two sediments, many will hesitate and consider the possibility that the answer will be "the same amount." We urge the students to discuss the question among themselves. Typically, some students will think about surface tension and suggest that smaller grains will retain more water. Again, we tally the student responses in columns on the board, then proceed to drain water from the second beaker of finer sediment. The volume of water yielded from the finer sediment is markedly smaller than that yielded from the coarser sediment, which indicates that the finer material has lower permeability. Typical values of permeability for common geologic materials are given in Table 8.2 in the textbook. A useful aside is to direct students to the textbook case study for Cape Cod (Box 8.3). The glacial sediments there have very high hydraulic conductivities, and contaminants from Otis Air Base are moving rapidly, at rates as high as half a meter per day. Because contaminants have traveled so far during the past half century, clean-up will be costly and difficult.

Demonstrating Darcy's Law

In a real aquifer, groundwater usually flows in response to gravitational forces. An example of flow from one spot in an aquifer to another was demonstrated when the tygon tubing was used to drain water from one beaker into another. The gradient of the tubing (change in height over distance) is actually the hydraulic gradient. It is possible to change the hydraulic gradient and to use this change to demonstrate Darcy's law. We ask the students how to make the water flow from the beaker at as fast a rate as possible. Most students realize that the greater the hydraulic gradient, the greater the flow rate of water into the empty beaker. We sometimes hold the two beakers against the blackboard and draw the hydraulic gradient onto the board, then use a stopwatch to clock the rate of flow. Using Darcy's law,

Velocity = Hydraulic conductivity × Hydraulic gradient

we can solve for hydraulic conductivity of the sediment in the beakers. A chart of typical values of hydraulic conductivity can be shown or referred to for comparison (see Table 8.2 in the textbook).

Establishing Links between Demonstrations and the Real World

Real groundwater behaves much like the water in this demonstration. Water fills void spaces between particles or along cracks. Water can be pumped from the ground if the pores are interconnected, but not all water can be removed because some remains behind, held to solid rock surfaces by surface tension. If an aquifer is contaminated (for example, by petroleum products from a leaking underground storage tank), geologists will attempt to clean up, or remediate, the aquifer. In one method of remediation, called pump and treat, the water and contaminants are pumped from wells. It is more difficult to extract contaminants from fine-grained aquifers because of the greater amount of surface area to which contaminants can become attached. Also, the remediation takes longer because of the low permeability of the aquifer and the low hydraulic conductivity of viscous contaminants. In addition, groundwater has recharge and discharge areas, places where new groundwater is added to or removed from the system.

Other Possible Demonstrations

The same beakers of sediment and water can be used during a discussion of groundwater contamination. Some contaminants, such as the salt used to de-ice roads in cold regions, are miscible; others, such as petroleum products, are immiscible. We commonly pour oil, such as WD-40 or transmission fluid, into one of our beakers to demonstrate the properties of a "floater," or an immiscible fluid that is less dense than water. If one tries to remove water from the beaker using tygon tubing inserted to the bottom of the beaker, the water table drops and the floater moves downward, coating many of the particles. If more water is flushed back into the beaker through the tubing, the oil floats upward again, but it takes some time for this to happen. Repeated rising and falling of the water table is common in the real world, and it explains the complex variation in the product that is found at sites of leaking underground storage tanks (see Figure 8.5 in the Slide Set).

Groundwater Budget for the Heavily Irrigated Central Valley, California

A water budget for the state of California demonstrates the changes associated with intensive use of groundwater, primarily for irrigation. In the state's arid Central Valley, groundwater has been pumped for many decades to irrigate fields that produce nearly half the fruits and vegetables in the United States. Before extensive pumping commenced, the underlying alluvial sand and gravel aquifers received 1.8 billion gallons per day (bgd) of recharge from precipitation and infiltration at the edges of the valley, along the foothills of adjacent mountains. This water moved down and toward the valley center, where it seeped into streams or was lost by evapotranspiration. The rate of groundwater discharge equaled the rate of recharge, and storage of water in the aquifers was fairly constant over the past few thousand years.

Data collected between 1961 and 1977 are representative of the present-day water balance for the valley. Since extensive pumping of more than 100,000 wells began in the early 20th century, 10.2 billion gallons of water have been removed from the ground each day, a rate five times greater than the rate of natural recharge! Much of this water (9.8 bgd) is returned to the ground from recharge, largely by infiltration of irrigation water derived from pumping and imported surface water. Because water levels have dropped hundreds of feet in some places, natural discharge has diminished to 0.3 bgd, less than 20 percent of its original rate. The sum of natural discharge and pumping in the Central Valley is about 0.7 bgd greater than the amount of recharge (outflows = 10.2 + 0.3 = 10.5 bgd; inflows = 9.8 bgd). As a consequence, 0.7 billion gallons of water are removed from storage each day, and the aquifer resource is depleted, or mined, over time.

The Atmospheric System

CHAPTER OBJECTIVES

Chapter 9 is similar to Chapters 6 through 8 in its organization. It begins with fundamental information about the atmosphere, climate, and weather; then examines atmospheric hazards, including tornadoes and hurricanes; then considers various types of atmospheric degradation, such as acid rain, smog, and ozone depletion; and concludes with environmental management of the atmosphere and cleanup technologies to control atmospheric emissions. Chapter 9 lays much of the foundation necessary to understand the content of the final two chapters of the book, both of which focus on environmental and climate change. The objectives of the chapter are to

Investigate the impact of the atmosphere's composition and structure on climate.

Examine the links between atmospheric circulation, regional distribution of moisture, and storms.

Discuss the role of human-induced atmospheric pollution in climatic and environmental change.

Chapter 9 opens with a review of recent severe weather conditions and poses the question of whether or not unusual events such as blizzards might result from human effects on the atmosphere. To assess this question, however, one must understand the structure and composition of the atmosphere and its interactions with other Earth systems, particularly life forms that alter atmospheric chemistry (such as plants that extract carbon dioxide from the atmosphere and release oxygen in return). Box 9.1 is devoted to summarizing an important global change in the atmosphere—an increase in oxygen content—resulting from the intertwined histories of atmospheric chemistry and life on Earth.

CHAPTER OUTLINE

1. The Atmosphere
   A. Present Atmospheric Composition
   B. Atmospheric Structure

2. Climate and Weather
   A. The Greenhouse Effect
   B. Differential Heating of Earth
   C. Tropospheric Circulation

3. Storms
   A. Development of Air Masses and Frontal Weather Systems
   B. Lightning and Thunder
   C. Tornadoes
   D. Tropical Storms and Hurricanes

4. Human Influence on Atmospheric Chemistry
   A. Acid Rain
   B. Smog
   C. Ozone Depletion

5. Air Pollution and Environmental Management
   A. Legislation
   B. Cleanup Technologies

SUGGESTED LECTURE OUTLINE

The headings listed below are related to topics addressed in Chapter 9 and provide an alternative structure by which to present the material during classes and/or discussion periods. Items marked with one asterisk are treated in the following section on suggested lecture and discussion topics; those marked with two asterisks are treated in the section on demonstrations and case studies.

The Atmosphere System
   Composition
   Changes in composition during Earth's history
   Structure from sea level to outer space

Climate and Weather
   Solar radiation, Earth's tilt, and seasons
   The greenhouse effect
   Climates on Mars and Venus (see also Chapter 2)
   Atmospheric circulation patterns

Atmospheric Hazards
    Tornadoes
    Hurricanes
    Blizzards

Air Pollution and Degradation
    The by-products of burning fossil fuels
    Acid rain**
    Smog
    Ozone depletion

Environmental Management of the Atmosphere
    The Montreal Protocol
    The Clean Air Act and its amendments
    Catalytic converters and scrubbers
    Pollution in rapidly developing countries such as China*

SUGGESTED LECTURE AND DISCUSSION TOPICS

Air Pollution in China

Another provocative topic for discussion is the issue of environmental degradation, particularly air pollution, associated with rapid industrialization in developing nations, including China and India. These countries fit the model of pollution overpopulation presented in Chapter 1 (pages 20–21). In this model, people use technologies that are grossly polluting, and the amount of pollution produced per unit of resource used is so high that extreme environmental degradation occurs.

During the Kyoto conference on global warming in December 1997, China made clear that its government will not jeopardize the country's recent rapid economic growth for the sake of minimizing possible global warming. Participants at the conference argued for global reductions in emissions of carbon dioxide gas into the atmosphere, because the gas might contribute to global warming. But China is still at a point that already developed nations occupied in the past, a point where economic growth is achieved at the expense of the environment. For developed nations like the United States and England, killing smogs that occurred in the 1940s and 1950s led to federal legislation to reduce emission of air pollutants. Thousands of residents of London, Pittsburgh, and other industrial cities died from respiratory ailments resulting largely from coal combustion and increasing use of automobiles in crowded cities.

Today, just more than a quarter of all deaths in China result from respiratory disease due largely to cigarette smoking and coal smoke. Lung cancer has increased nearly 20 percent in major Chinese cities since 1988, where the air is ranked as some of the filthiest on Earth. At least five of China's cities are on the world's list of worst-polluted. With nearly one-fourth of the world's populace, immense reserves of coal, and escalating consumption of this dirtiest of fossil fuels, China's problems are of importance to the world. Yet just as its predecessors chose to deal with immediate, local problems of air pollution first, China prefers to postpone worrying about global warming before treating its own problems of filthy air and acid rain.

What the Chinese are concerned with at the moment, besides ensuring that their booming economy continues to boom, is soot and dust, or total suspended particulates (TSP). Levels of TSP are four to nine times higher than the guidelines set by the World Health Organization, which are 60 to 90 micrograms per cubic meter per year. For comparison, TSP values range from 40 to 60 in most American cities but are as high as 400 to 800 in the wintertime in cities in northern China (see Figure 11.12 in the textbook).

Chinese officials struggle to deal with the problems of environmental degradation because so many of their people still have insufficient heat, low-paying jobs, and miserable living conditions. They argue that regulating or penalizing polluters is bad for economic growth. And yet the costs of pollution are high: the damage to forests and agriculture from acid rain is $2.8 billion per year, and health-care costs are increased due to the number of respiratory ailments. Estimates of the cost of environmental degradation range from 7 to 15 percent of the gross domestic product (see "Our Real China Problem" by Mark Hertsgaard, published in the November 1997 *The Atlantic Monthly*, pages 96–114).

If this information is presented to students, it is easy to generate animated discussions about what should be done to minimize the threat to the global environment, for China's pollution does not respect national borders. Do already-industrialized nations have the right to force China to curb its pollution at a point in its economic growth that is earlier than when those finger-pointing nations curbed their own emissions?

DEMONSTRATIONS AND CASE STUDIES

Environmental Problem Solving: Acid Rain Deposition

This problem was modified substantially from three problems in Harte (1988), *Consider a Spherical Cow: A Course in Environmental Problem Solving*. Mill Valley, Calif.: University Science Books.

Problem Background

Acid rain results from the formation of sulfuric acid in the atmosphere. Sulfuric acid forms from sulfur dioxide (SO_2), sulfate (SO_4^{-2}), and hydrogen sulfide (H_2S) emitted from a variety of natural and anthropogenic sources. In the atmosphere, sulfur dioxide oxidizes to form sulfate, which in turn dissolves to form sulfuric acid, as follows.

Oxidation of sulfur dioxide to form the sulfate ion:

$$SO_2 + O_2 \rightarrow SO_4^{-2}$$

Reaction of the sulfate ion with hydrogen to form sulfuric acid:

$$SO_4^{-2} + H_2 \rightarrow H_2SO_4$$

This moderately strong acid is then deposited on the Earth's surface in solution as snow or rain and enters the biosphere.

Our atmosphere has contained a background stock of sulfur (S) from natural sources for billions of years, but the burning of fossil fuels since the Industrial Revolution has increased the amount of sulfur in the atmosphere by about 85 percent. The globally averaged concentration of sulfur has increased from about 0.2 ppb(v) (parts per billion by volume) to 0.37 ppb(v). Although this increase alone would result in more acidic precipitation, the greater problem is that anthropogenic sources of sulfur are highly concentrated regionally in areas such as the northeastern United States, where emissions from smokestacks and automobiles are high. Another problem is that the residence time of sulfur in the atmosphere is very short. As a consequence, parts of the globe close to sources of sulfur have very acidic precipitation. Even worse, regions that are downwind from the sources have the most acidic precipitation, even if those areas don't have their own sources of sulfur emissions. Canadians who live across the border and downwind from the industrial belt in the northeastern United States know this phenomenon well.

Problem

Determine the residence time of sulfur in the form of sulfur dioxide (SO_2) in the atmosphere. Then use this residence time to estimate how large an area of the northeastern United States will be affected by emissions from sources extending westward from Bethlehem, Pennsylvania, to the Ohio Valley just west of Pittsburgh, Pennsylvania.

Solution

Sulfur is added to the atmosphere from both natural and anthropogenic sources. Natural sources and their annual flows (in metric tons of sulfur; 1 metric ton equals 1000 kg) are:

Source of Sulfur	Annual Flow (metric tons)
Volcanic emissions	0.30×10^8
Sea spray	0.38×10^8
Biomass	0.82×10^8
Total background sulfur	1.5×10^8

Sulfur from these natural, or background, sources is added to the atmosphere in one of three forms: H_2S (hydrogen sulfide), SO_2 (sulfur dioxide), or SO_4^{-2} (sulfate). Hydrogen sulfide converts to sulfur dioxide very quickly in the atmosphere, but sulfate does not. About 66 percent of the background sulfur emissions, or 1.0×10^8 tons of sulfur per year, occur as hydrogen sulfide and sulfur dioxide; these emissions can be treated as sulfur dioxide because of the rapid conversion of hydrogen sulfide to sulfur dioxide.

So, our background flow rate for sulfur as SO_2 is 1.0×10^8 tons of sulfur per year. We know from measurements in remote, nonindustrial parts of the world that the background, or natural, stock of sulfur in atmospheric SO_2 is 1.15×10^6 tons. Therefore, we can calculate a residence time for sulfur in atmospheric SO_2 as follows:

$$\frac{\text{Stock of S in the atmosphere as } SO_2}{\text{Inflow of S as } SO_2 \text{ into the atmosphere}} =$$

$$\frac{(1.15 \times 10^6 \text{ tons S})}{(1.0 \times 10^8 \text{ tons S})/\text{yr}} =$$

$$0.015 \text{ yr} = 4.2 \text{ days}$$

Recently, combustion of fossil fuels and smelting of metal ores has resulted in a substantial increase in the atmospheric stock of sulfur. In 1980, anthropogenic emissions amounted to 0.85×10^8 tons of sulfur per year, or 85 percent of the natural background rate of emissions from volcanoes, sea spray, and biomass. An astonishing 12 percent of this is emitted from only one airshed, the industrial belt in the northeastern United States, which makes up only 0.2 percent of Earth's surface! This belt extends from the Ohio Valley in the west to the Atlantic coast in the east and from the latitude of Washington, D.C., northward as far as the Canadian border.

The northeastern United States can be treated roughly as a reservoir, or airshed, for emissions of sulfur. Treating it as a reservoir that contains a stock of sulfur is somewhat erroneous because (1) its boundaries are crude, (2) emissions from outside the region can enter it, and (3) emissions from within the region can leave it. Although these and other complicating factors, such as weather conditions and the composition of atmospheric gases, affect the amount of time a given SO_2 molecule resides in the atmosphere, we can estimate an approximate distance that such a molecule might travel from its source. This value can then be viewed as the downwind distance over which acid deposition might occur from the source of emissions.

If the Ohio-Pittsburgh-Bethlehem area is the source of the SO_2, we need to know an average wind speed and direction for this region. From meteorological data, we find that the annual average wind speed is 15 km/hr, and the mean direction of wind flow is N25°E. We determined that the average residence time of a typical SO_2 molecule in the atmosphere is 4.2 days, or 100.8 hours (4.2 days × 24 hours/day = 100.8 hours). Distance traveled is velocity multiplied by time. So, a typical sulfur dioxide molecule might travel a distance of

$$15 \text{ km/hr} \times 100.8 \text{ hr} = 1512 \text{ km}$$

This calculation tells us that a typical molecule of sulfur dioxide from the Ohio Valley might be able to travel as far as, or even just beyond, the Canadian border. We can imagine a 1512-km plume of emission products extending downwind from the emission sources, with the greatest concentrations of products such as sulfur dioxide closer to the source and decreasing concentrations towards the downwind tail of the plume. Imagining such a plume, and considering the short residence time of sulfur dioxide, makes it clear why lakes in the Adirondacks are being acidified and Canadian and New England maple forests are severely stressed and dying, even though no major sources of emissions exist in these regions.

CHAPTER 10

The Ocean and Coastal System

CHAPTER OBJECTIVES

Chapter 10 is the last chapter in Part III, Fluid Earth Systems. In some ways, the oceans deserve more than just a single chapter. Groundwater and surface water each get a chapter, but far more water is contained in the oceans, and the oceans play a number of critical roles in the Earth system. Oceans modulate Earth's temperature, cycle carbon from the atmosphere and surface water to the deep ocean and rock reservoirs, and control the position of the world's coastlines in response to the changing level of the ocean surface.

This textbook focuses on environmental geology, using Earth system science as a perspective from which to view and understand important environmental issues. For this reason, we highlight those aspects of physical, chemical, and biological oceanography that are relevant to such hazards as sea level change and coastal erosion and to such environmental issues as ocean pollution and coastal development. Many aspects of the oceans and ocean basins are presented in the context of environmental issues. For example, bathymetry and seafloor features are discussed in the context of the Law of the Sea and the national and international attempts to divide up the oceans and their resources.

Like Chapters 5 through 9, Chapter 10 begins with a section on fundamental processes and materials, focusing on the composition and circulation of seawater. This section is followed by a discussion of ocean and coastal hazards, including sea level rise, coastal erosion, and tsunamis. The final section treats ocean pollution and recent efforts to protect the world's oceans.

The last major cycle treated in the textbook—the carbon cycle—is presented in this chapter. Like another example in this chapter, a box on El Niño climatic events (Box 10.2), the carbon cycle could just as easily have been included in the atmosphere chapter (Chapter 9). In both cases, the phenomena discussed are related to the linked atmosphere-ocean Earth systems. We chose to put the carbon cycle and the El Niño discussion in the chapter on oceans because it follows the discussion of the atmosphere, and the students will now be prepared to understand both systems and how they are linked after having read Chapters 9 and 10.

Specific objectives of this chapter are to

Examine the nature of ocean basins and their shoreline boundaries.

Analyze global ocean circulation and its links to other Earth systems.

Investigate the causes of coastal erosion, sudden flooding, and rise of sea level.

Evaluate the effects of humans on the seas and efforts to control marine pollution.

CHAPTER OUTLINE

1.   The Ocean Basins
      A.   Features of the Seafloor
      B.   Ocean Basins, Coastlines, and Plate Tectonics

2.   Seawater Properties and Ocean Circulation
      A.   The Salinity of Seawater
      B.   The Ocean Reservoir and Flux of Salt Ions
      C.   Nutrients in Seawater
      D.   The Carbon Cycle
      E.   Ocean Circulation

3.   Hazards of the Sea
      A.   Changes in Sea Level
      B.   Coastal Erosion
      C.   Storm Surges, Flooding, and Delta Erosion
      D.   Tsunamis

4.   Ocean Pollution
      A.   Effects of Population Growth and Development
      B.   Waste Disposal and Runoff along Continental Shelves
      C.   Wetlands and Their Role in Protecting Coasts
      D.   Pollution in the Deep Ocean
      E.   Protecting the Oceans

SUGGESTED LECTURE OUTLINE

The headings listed below are related to topics addressed in Chapter 10 and provide an alternative structure by which to present the material during classes and/or discussion periods. Items marked with one asterisk are treated in the following section on suggested lecture and discussion topics; those marked with two asterisks are treated in the section on demonstrations and case studies.

Plate Tectonics and the Ocean Basins
    Mid-ocean ridges, deep-sea trenches, and transform faults
    Passive margins, active margins, and continental shelves
    Abyssal plains and seamounts
    Hot spots and volcanic islands (see also Box 4.3)

Ocean Exploration
    Early bathymetric studies (HMS *Challenger*)
    Sonar studies: side-scan and multibeam
    Satellite altimetry
    Underwater submersibles

Seawater Chemistry
    The ocean reservoir and salt flux
    Salinity of seawater
    Nutrients in seawater

Ocean Circulation
    The surface mixed layer
    The deep ocean

Carbon Cycling and the Linked Atmosphere-Ocean Systems
    The natural carbon cycle
    Human impact on the carbon cycle
    The mystery of the missing carbon*

Oceans and Climate Change
    El Niño climatic events

Sea Level Change
    Causes of changing sea level**
    Evidence of sea level rise in the 20th century

Coastal Landforms and Longshore Currents
    Barrier islands

Coastal Erosion
    Groins and jetties
    Seawalls

Ocean Pollution
    Ocean dumping
    Continental runoff
    Dredging

SUGGESTED LECTURE AND DISCUSSION TOPICS

The Missing Carbon

The carbon cycle is presented on pages 301–304 of the textbook and is accompanied by two related illustrations. Figure 10.12 is the carbon budget and—like the figures that illustrated the rock, water, and copper budgets in earlier chapters (Figures 2.19, 2.23, and 5.10, respectively)—this figure illustrates the various carbon reservoirs and fluxes as drawn to scale relative to one another. Figure 10.13 illustrates the more conventional and conceptual view of the carbon cycle.

It is very helpful to point out to students that a similar pairing of illustrations can be found for the rock cycle (Figures 2.19 and 4.11) and the hydrologic cycle (Figures 2.21 and 2.23). Make clear to them that the scaled versions of the systems views of the rock, hydrologic, copper, and carbon cycles contain quantitative information from which one can derive residence times or evaluate the magnitude of human impact on Earth cycles. Residence times for water were calculated in Chapters 2 and 7, and for copper in Chapter 5.

In Chapter 10, we examine the human impact on the carbon cycle in a special cycling feature marked with an arrow that begins on page 303 and ends on 304. This example focuses on the so-called missing carbon. In our classes, we present the missing carbon as a state-of-the-art scientific problem, one that is slowly being solved as different pieces of information are obtained from a variety of types of studies. We developed Figure 10.14 as a way of presenting the problem and illustrating why scientists think that some carbon is missing. This illustration is a carbon budget just for the amount of carbon released into the atmosphere by human activities (burning fossil fuels and forests).

Our experience has been that students rarely understand Figure 10.14 unless we present its individual components one at a time and make frequent reference to how it differs from the natural carbon cycle. We present all the known parts and conclude by noting that some carbon added to the atmosphere is missing—that is, it is not stored in the atmosphere and we don't know exactly where it goes. We can conclude, however, that it must go back to the biosphere, for we are fairly certain of how much goes into the oceans.

DEMONSTRATIONS AND CASE STUDIES

Environmental Problem Solving: Changing Ice Volumes and Sea Level

This material can be used in whole or in part for class discussions to illustrate why so many scientists are concerned about global warming. It can also be used as part of a lab exercise. One semester, we used these problems along with topographic maps of major coastal cities (such as Pearl Harbor, Miami, and San Francisco). We had students identify areas that would be inundated under various scenarios of sea-level rise. The exercise was very successful and proved to be a good way to explain the concept of a contour line. The following exercises estimate how much sea level would rise (1) if all glacial masses on Earth were to melt, and (2) if just the West Antarctic ice sheet were to melt.

Problem Background

The volume of water on Earth has been constant for billions of years, but it is continuously exchanged between one reservoir and another, and its distribution in these reservoirs varies over time. Ice on the continents and water in the oceans are dynamically linked and form the principal exchange of water on Earth. The melting of 35–50 km^3 of ice and the drainage of all this water from continents resulted in a substantial increase in the stock of water in the oceans approximately 18,000 years ago; in fact, these were the main causes of the approximately 120-m rise in sea level known to have occurred from dating of submerged coral reefs in Barbados (see Figure 10.22 in the textbook). How much stock was added to the ocean basins, and is this amount sufficient to account for the rise in sea level determined independently?

Let's imagine that it is 18,000 years ago and calculate how much sea level will rise if we melt 35–50 km^3 of ice. Then let's compare this number with the estimate of sea-level rise from studies of submerged coral. This will be problem 1. Next, in problem 2, let's determine how much higher sea level could rise above the present level if we were to melt the current Antarctic and Greenland ice sheets.

Problem 1: Rising Sea Level due to Melting of the World's Ice Sheets

Assumptions and Background Steps

Changes in Volume When Converting Ice to Water

To determine how much sea level would rise as a result of melting 35–50 million km^3 of ice, we need to calculate the volume of water that would result from melting. Ice is less dense than water, which is why icebergs float in the oceans. Ice has a density (density = mass/volume) that is 0.9 times (90 percent) that of water. Its density is 0.9 g/cm^3, or 0.9×10^9 kg/km^3. Liquid water has a density of 1.0 g/cm^3 (1.0×10^9 km^3). When one cubic kilometer of ice melts, it will decrease in volume by approximately 10 percent:

$$1 \ km^3 \ ice = 0.9 \times 10^9 \ kg \ H_2O \times 1 \ km^3 \ H_2O = 0.9 \ km^3 \ H_2O$$

$$1 \ km^3 \ ice = 1 \times 10^9 \ kg \ H_2O$$

If 35,000,000 km^3 of ice melted, the volume of water released would be

$$35,000,000 \ km^3 \ ice = (35,000)(0.9 \ km^3 \ water) = 31,500,000 \ km^3 \ water$$

Melting of the maximum amount of ice estimated, 50,000,000 km^3, would result in 45,000,000 km^3 of water.

Determining the Surface Area of the World's Oceans During Full-Glacial Times

The surface area of the world's oceans today is 361,300,000 km^2, but during the last ice age, when sea level was about 100–120 m lower, its surface area was less. The amount of decrease was about equal to the amount of additional land area that was exposed at that time. One way to obtain this amount is to determine the amount of land area that is between 0 and −100 m in altitude. Bathymetric studies of the ocean floor have been done all over the world, and from them we have learned that 8,000,000 km^2 of land is between 0 and −100 m in altitude. If we subtract this area from the present area of the world's oceans, we have a rough estimate of the area of the oceans (A_O) during the last full-glacial low stand of sea level:

$$A_O = 361,300,000 \text{ km}^2 - 8,000,000 \text{ km}^2 = 353,300,000 \text{ km}^2$$

Solution

There are many ways to determine how much sea level would rise if the continental ice sheets melted, releasing 31,500,000 to 45,000,000 km^3 of water. First, let's assume that all meltwater would drain into the oceans, rather than recharging groundwater or increasing atmospheric moisture. Second, let's assume that all rise in sea level would be solely a result of an increase in the mass of water, rather than an increase in volume due to global warming and thermal expansion of the water in the oceans. We will discuss the sources of error in these assumptions later.

We present a simple solution in which we will assume that the world's oceans fill one basin with vertical sides. This is a quick, back-of-the-envelope type of solution. This solution is inaccurate, however, because we know that the coastal plains flanking most continental margins (for example, the Atlantic seaboard) are gently sloping. Despite this source of error, this simple solution gives a ballpark estimate of the magnitude of sea-level rise associated with different amounts of ice melting.

We know how much water is pouring into the ocean basin, and we have assumed that the shape of the basin is a rectangular prism. The new volume of water divided by the surface area it can cover gives us the height of the new water column. For the minimum estimate of meltwater, 31.5 million km^3:

$$\frac{\text{Volume}}{\text{Area}} = \frac{31,500,000 \text{ km}^3}{353,300,000 \text{ km}^2} = 0.09 \text{ km} = 9 \text{ m}$$

For the maximum estimate of meltwater, 45 million km^3, the sea-level rise would be 0.1 km, or 100 m. These estimates are surprisingly close to the value of sea-level rise estimated from dating of submerged coral reefs in Barbados (121 ± 5 m).

Problem 2: Rising Sea Level due to Global Warming

Background

Based on the results just obtained, we can now determine what would happen to the global sea level if the modern ice sheets were to melt. We know that a back-of-the-envelope calculation as in the solution to problem 1 is adequate for our purposes. In fact, it differs from a solution that considers the slope of the continental shelves by only a few centimeters. Many scientists have warned that global warming might result in rapid melting of the West Antarctic ice sheet. This ice sheet is grounded on bedrock that is mostly below present sea level and is flanked by numerous floating ice shelves. Consequently, some scientists consider it to be unstable and prone to surging, or high-speed flow, into the Antarctic Ocean as a result of small increases in temperature. The volume of the West Antarctic ice sheet is 3,672,000 km^3. The volume of the much larger East Antarctic ice sheet is 17,928,000 km^3.

Solution

Let's assume the unusual circumstance that all of the West Antarctic ice sheet would melt and assume that the sides of the ocean basins are vertical, as above. Again, we must first convert the amount of ice melted to a water equivalent that considers the difference in density of ice and water:

$$V_T = 3{,}672{,}000 \text{ km}^3 \text{ ice} \times \frac{0.9 \text{ km}^2 \text{ water}}{1 \text{ km}^3 \text{ ice}} = 3{,}304{,}800 \text{ km}^3 \text{ water}$$

The height to which the water column will rise is the total volume of water divided by the area it covers:

$$y = V_T/A_O,$$

and A_O at present is 361,300,000 km^2, so $y = 9$ m. A 9-m rise in sea level would inundate many of the world's coastal cities, including parts of Miami, Ho Chi Minh City, Bangkok, and New Orleans.

If all the ice sheets in the world (both East and West Antarctic ice sheets and the Greenland ice sheet) were to melt, the total rise in sea level would be 66 m, since their total volume is 23,940,000 km^3 of water equivalent. Although such an occurrence is unlikely considering the prolonged cooling of global temperatures that has been occurring for the past 50–60 million years, no-ice conditions did exist for much of the time between 15 and 65 million years ago, during the early Tertiary period, and sea level was much higher than at present. Furthermore, only 125,000 years ago, during the last major interglacial period, sea level is thought to have been 6 m higher than at present, and some scientists have proposed that the cause of the higher stand might have been melting of the West Antarctic ice sheet. Considering our estimate of a 9-m rise in sea level due to melting of this ice sheet, one can see why scientists might have developed such a hypothesis.

Sources of Error

There are a number of sources of error in the solutions detailed above. We mentioned the difficulty of determining the perimeter of the ocean basins, the surface area of the oceans during the last low stand, and the slopes of the continental shelves. More important sources of error, however, probably result from the temperature increases that initially melted the ice sheets. Increased global temperatures would result in warming of the ocean and consequently in thermal expansion of ocean water. In that case, no change in the mass of water would occur, but a change in volume would. Such a cause of rise in sea level is called steric, and it is estimated to cause ~1 m of sea level rise per degree Celsius increase in temperature throughout the ocean water column. Some scientists invoke this source of sea-level rise to explain the unusually high early Tertiary and Cretaceous sea-level stands.

Another source of error in our calculations is that we did not consider the isostatic effects of loading such a large mass of water into the ocean basins. Because the ocean floor consists of somewhat buoyant lithosphere floating upon plastic asthenosphere, the floor would be depressed to compensate for the increased weight. Because the Earth's lithosphere is elastic (it can be treated as an elastic beam), loading the ocean floor with water and unloading ice from the continents would result in uplift of the continents as well as subsidence of the ocean floor. Recent deglaciation might have resulted in up to 8 m of depression of the ocean floor and 16 m of uplift in some continental areas. Isostasy would affect the estimates of sea level rise due to melting.

Energy and the Environment

CHAPTER OBJECTIVES

Chapter 11 is the first of three in Part IV, Energy and Change in Earth Systems. The chapter begins with an overview of Earth's energy system that is designed to serve as a review of material presented on pages 47–50 of Chapter 2. The remainder of the chapter, with the exception of the last section, treats six major categories of energy resources: petroleum, other fluid hydrocarbons, coal, geothermal, nuclear, and solar. The final section of the chapter treats energy efficiency and conservation from the perspective of fundamental laws of thermodynamics.

Although the beginning of the chapter emphasizes that the majority of energy on Earth comes from the Sun, the current global consumption of energy resources for human activities relies heavily on fossil fuels, particularly petroleum and coal. The chapter sections for these fossil fuels are organized similarly, beginning with the geologic processes that produce the fuels, then the occurrence and distribution of the substances, followed by a discussion of the threat of resource depletion, and concluding with an assessment of the environmental impact of consumption of each fuel.

Tar sands, oil shales, and geothermal energy resources are quite interesting in a geologic sense, but they contribute very minor amounts to the global energy mix, with a few local exceptions (for example, geothermal energy in Iceland). For this reason, the sections on these resources are short.

Several energy resources that are closely linked to processes resulting from solar energy and the cycling of solar energy on Earth are treated in the solar energy section. These resources include wind, hydroelectric, and biomass energy. In reality, even fossil fuels could be treated in this section, because they are the result of solar energy that has been converted to chemical energy and buried as hydrocarbons.

The overall goal of this chapter is to illustrate how the cycling of energy on Earth contributes to the formation of certain types of energy resources that can be tapped for human consumption. Specific objectives of this chapter are to

Trace flows of energy through Earth systems.

Investigate responsible ways to recover, use, and conserve nonrenewable fossil fuels.

Confront the risks involved in producing nuclear energy and disposing of nuclear waste.

Discover why solar radiation is the only perpetual source of energy safely available worldwide.

CHAPTER OUTLINE

1.   Earth's Energy System
     A.   Sources of Energy on Earth
     B.   Energy Transfers and Photosynthesis
     C.   Energy Resources

2.   Petroleum
     A.   Origin of Petroleum
     B.   Petroleum Traps
     C.   Finding, Extracting, and Refining Petroleum
     D.   Will We Run Out of Petroleum?
     E.   Environmental Impact of Using Petroleum

3.   Other Fluid Hydrocarbons
     A.   Tar Sands
     B.   Oil Shales

4.   Coal
     A.   Origin of Coal
     B.   Global Distribution of Coal Deposits
     C.   Will We Run Out of Coal?
     D.   Environmental Impact of Using Coal

5.   Geothermal Energy

6.   Nuclear Energy
     A.   Fission and Fusion
     B.   Problems with Nuclear Energy

7.   Solar Energy
     A.   The Potential of Solar Energy
     B.   Wind Energy from the Sun
     C.   Water Energy from the Sun
     D.   Biomass Energy

8.   Energy Efficiency and Conservation

SUGGESTED LECTURE OUTLINE

The headings listed below are related to topics addressed in Chapter 11 and provide an alternative structure by which to present the material during classes and/or discussion periods. Items marked with one asterisk are treated in the following section on suggested lecture and discussion topics; those marked with two asterisks are treated in the section on demonstrations and case studies.

Earth's Energy System
    Earth's energy budget (refer to Figure 2.16)*
    Flows of energy from one Earth system to another

Energy Resources
    Fossil fuels
        Petroleum
        Coal
        Tar sands and oil shales
    Nuclear energy
    Solar energy
    Hydroelectric energy
    Wind energy
    Biomass
    Geothermal energy

Environmental Impact of Using Different Energy Resources
    Fossil fuels
        Smog
        Acid rain
        Greenhouse gases*
    Nuclear power and production of radioactive wastes
    Hydroelectric dams
        Effects on migration and spawning of fish
        Effects on sediment loads in streams
        Effects on stream discharge and channel patterns

SUGGESTED LECTURE AND DISCUSSION TOPICS

Earth's Energy System, Energy Resources, and Change

As noted earlier in this manual, students benefit from instructors' help in making connections between content presented in different chapters of their textbooks or in various lectures throughout the semester. It is worth reminding students that Earth's energy system was introduced briefly in Chapter 2, on pages 47–50. Earth's energy budget is illustrated in this same section, in Figure 2.16. The first major section of Chapter 11 is a discussion of the transfer of energy from one Earth system to another via the major chemical reactions of photosynthesis, respiration, decomposition, and combustion. These reactions are important to a fundamental understanding of the cycling of solar energy through Earth systems and the formation of fossil fuels. The combustion reaction is particularly important to discussions of the environmental impact of burning fossil fuels, as discussed in the following suggestion for a lecture/discussion topic.

Energy is of importance to environmental geology for two primary reasons. First, human population growth has been associated with a similar pattern of growth in energy consumption. New technology that results in even greater consumption of energy products—such as invention of the internal combustion engine—has fueled major societal changes that in turn have led to even greater rates of population growth. From a geologic perspective, these changes are important because many of the energy resources are associated with geologic deposits and are found by geologists. From an environmental perspective, they are important because of the impact that their use has on Earth systems.

Energy is important to environmental geology for a second reason, one that is quite different from an interest in energy resources and environmental degradation. This second reason is the role of energy in the changes that occur in Earth systems. When the amount of energy reaching Earth's surface changes, so too do processes, rates of processes, and environments on Earth. Energy drives all processes on Earth, and any deviations in the amount of energy flowing into or out of a given system result in changes in that system. For this reason, the last two chapters of the book are about changes in Earth systems. Environmental change, and particularly climatic change, is of increasing importance in modern environmental geology and distinguishes it from conventional environmental geology, which focuses largely on hazards and resources.

By-Products of Burning Fossil Fuels

When discussing the environmental impact of burning fossil fuels for energy, it is valuable to get students to think about their own contributions to atmospheric pollution. Few people seem fully aware of the volume of waste products emitted to the atmosphere by automobiles. One need only stand near the tailpipe of a car with its engine running to be vividly reminded that automobiles cause pollution. As noted on page 285 of the textbook, nitrogen oxide emissions from cars have decreased 76 to 96 percent since 1967, but the total number of people and cars has increased equally remarkably, so that the volume of atmospheric emissions from automobiles has merely stabilized or slowed in its rate of increase.

When petroleum is burned, the carbon and hydrogen combine with oxygen from the atmosphere to form carbon dioxide and water vapor. A butane lighter can be used in class to demonstrate this reaction. Butane is a distillation product of crude oil and contains about 2.5 atoms of hydrogen per atom of carbon. When a butane lighter is flicked, oxygen from the atmosphere combines with the hydrocarbon, and combustion between the two is aided by a spark. Although students will see only a flame, the instructor can point out that the following reaction occurs, producing carbon dioxide, water vapor, and energy (heat).

$$C_4H_{10} + 6.5\,O_2 \rightarrow 4\,CO_2 + 5\,H_2O + energy$$

A typical American car emits 22 pounds of carbon dioxide per gallon of fuel burned, or 1.12 pounds of carbon dioxide per mile for a passenger car. The average car in the United States is driven about 10,000 per year, and some 145 million cars are on the road. Using these statistics to calculate the amount of carbon dioxide emitted to the atmosphere from cars yields an annual value of about 0.8 billion tons. The total amount of carbon dioxide emitted to the atmosphere from burning fossil fuels in the United States is 5.5 billion tons each year, so automobiles contribute about 15% of the annual loading of carbon dioxide into the atmosphere.

Understanding, Tracing, and Predicting Environmental Change

CHAPTER OBJECTIVES

One of the most important objectives of an introductory course in geology is to imbue students with a deep awareness of the fact that the Earth system changes over time. Another equally important aspect of geology is an understanding of time scale. If one is asked, for example, whether or not global warming is occurring, a geologically literate person should answer with the question, "Over what time scale?" The Earth might be warming over the past few hundred years, but certainly not over the past few tens of millions.

In Chapters 12 and 13, we end this textbook in environmental geology by focusing on what is known about environmental change and the time scales over which it occurs:

How does Earth's climate compare with the climates of Mars and Venus and why are these planets' climates so different (Chapter 12)?

What causes environmental changes on Earth (Chapter 12)?

What is the evidence for environmental change in the geologic and biologic records (Chapter 12)?

What is the history of Earth's environments and climates since its origin (Chapter 13)?

How have Earth's environments and climates varied since upright-walking primates evolved, and have they affected human evolution (Chapter 13)?

What types and magnitudes of environmental change are occurring as a result of recent human activities (Chapter 13)?

Addressing all these questions fairly comprehensively is too large a task for a single chapter, so we chose to close the textbook with one chapter on the causes and indicators of environmental change (Chapter 12) and another on the history of environmental change on Earth (Chapter 13). We have combined these chapters for the purposes of this manual.

Specific objectives of Chapter 12 are to

Examine the origin of Earth's climate.

Analyze some of the causes of climatic and environmental change.

Discover how proxies such as sediments, tree rings, pollen grains, and glacial ice are used to unravel Earth's history of environmental change.

Specific objectives of Chapter 13 are to

Place our modern environment within the context of Earth system processes and changes that occurred billions, millions, and thousands of years ago.

Explore the relationship between environmental change and human evolution.

Examine how modern human activities can affect the environment.

CHAPTER OUTLINES

Chapter 12

1.   Climate on Terrestrial Planets
      A.   Mars: A Frozen Planet
      B.   Venus: A Runaway Greenhouse Effect

2.   Causes of Climate Change
      A.   Influence of Plate Tectonics on Climate
      B.   Influence of the Oceans on Climate
      C.   Influence of Earth's Orbital Parameters on Climate
      D.   Climatic Feedbacks

3.   Indicators of Environmental Change
      A.   Geologic Records of Climate and Environment
      B.   Biologic Records of Climate and Environment

Chapter 13

1.   Earth before Humans
        A.   Hothouse Earth
        B.   Early Episodes of Global Cooling
        C.   Climatic Impact of Plate Tectonics
        D.   Meteorite Impacts and Environmental Change
        E.   The Last Great Ice Age

2.   Humans and Environmental Change
        A.   Aridification of Africa and Primate Evolution
        B.   The Holocene Epoch and Human History
        C.   The Medieval Warm Period
        D.   The Little Ice Age

3.   Humans as Agents of Environmental Change
        A.   Global Warming
        B.   Global Metal Pollution
        C.   Global Land Transformation

SUGGESTED LECTURE OUTLINE

The headings listed below are related to topics addressed in Chapters 12 and 13 and provide an alternative structure by which to present the material during classes and/or discussion periods. Items marked with one asterisk are treated in the following section on suggested lecture and discussion topics; those marked with two asterisks are treated in the section on demonstrations and case studies.

How Has Earth's Climate Varied over Different Time Scales?*
        Billions of years: ice ages and hothouse conditions
        The past 60 million years: global cooling
        The past few million years: glacials and interglacials of the last great ice age
        The past few thousand years: warm periods and little ice ages
        The past few centuries: Medieval Warm Period, Little Ice Age, and global warming

What Causes Climate Change over Different Time Scales?*
        Billions of years: drifting continents and volcanic activity
        Episodic change from meteorite impacts
        The past 60 million years: uplifting mountain belts and changing ocean currents
        The past few million years: Earth's orbital parameters
        The past few thousand years: climatic feedbacks
        The past few centuries: greenhouse gases, the Industrial Revolution, deforestation

Evidence of Environmental Change
    Glacial erosion and deposits
    Sedimentary deposits
    Paleo-lake shorelines
    Ancient forests and changing water tables**
    Ice core studies
    Fossils, middens, and tree rings
    Oxygen isotopes in marine fossils

Global Warming and Greenhouse Gases
    Is global warming occurring?
    Should we reduce greenhouse gas emissions with an international treaty?

SUGGESTED LECTURE AND DISCUSSION TOPICS

Climate Changes over Different Time Scales and Their Causes

Three figures can be presented (as overhead transparencies) in sequence to illustrate the types of climate changes that have occurred in Earth's history over different time periods. As these figures are presented, the instructor can explain possible causes of climatic change on those time scales.

Either Figure 3.10 or Figure 13.2 is useful for discussing climatic change over periods of billions of years. Of importance in these figures are the three periods of hothouse conditions (shown in red) and the five ice ages (shown in blue) that mark Earth's history. Four of Earth's known ice ages have occurred in the past billion years, and Earth still is experiencing the fifth ice age. The types of change on Earth that might have contributed to fluctuations between hothouse and icehouse conditions are those that operate over long time scales, such as the drifting of continents from low to high latitudes and changes in the rate of seafloor spreading (see pages 368–369 in the textbook).

It is also important to point out in Figure 3.10 the global cooling that has occurred over the past 65 million years. Coincident with this cooling has been the collision of India with Asia (see page 368 and Box 6.1), and some scientists suggest that uplift associated with collision might have increased rates of chemical weathering of rocks and resulted in a drawdown of carbon dioxide from Earth's atmosphere. As a consequence, Earth has cooled over a period of tens of millions of years. Other possible causes of climate change on this time scale are the opening and closing of landmasses and their resultant effect on oceanic currents (see the discussion in Chapter 2 of this manual). It is also important to note the concurrence of the evolution of the human species with the approximate onset of the most recent ice age. Possible correlations between these phenomena are discussed on pages 398–400.

Figure 12.7 illustrates the effects of various orbital parameters on the amount of summer radiation received on Earth (the Milankovitch cycles). Part (b) of this figure shows Earth's fluctuation from full glacial to interglacial conditions over the past 600,000 years. All of this time, and in fact the 2 million years or so before that, are considered to be within Earth's fifth and most recent ice age, but within this icehouse period the Earth's climate has warmed and cooled. Important points to mention about this figure are the cyclicity of the warm and cool periods and the similarities in maximum cooling and warming.

Figure 13.18 illustrates changes in Earth's temperature for the past 150,000 years. This figure is detailed enough to discuss differences between the penultimate interglacial 124,000 years ago and the present interglacial: the penultimate interglacial might have been warmer by a few degrees Celsius. It also provides enough detail to discuss possible causes of minor fluctuations (the interstadial episodes). Possible causes of climate change on the scale of thousands to tens of thousands of years are varying orbital parameters, climatic feedbacks, volcanic eruptions, changes in ocean current patterns, and human activities.

We like to close a discussion of the different scales of environmental change with an attempt to predict the future. In their book *Ice Ages* (1979, Enslow Press), John and Katharine Imbrie have written about the possibility of a superinterglacial period that might be brought upon ourselves as a result of emission of greenhouse gases and deforestation. The instructor can sketch such possible warming on Figure 13.18 and ask the students whether or not they think that global warming will be good or bad for the world's populace and economy. This question can be followed with a discussion of the recent international climate conference held in Kyoto, Japan, and the treaty that resulted. Should the world's nations attempt to reduce emission of greenhouse gases? This issue is already stirring up the U.S. Congress and is sure to be raised during the next presidential election. Much material is available on the Internet about the climate treaty, and students could be asked to do a short reading about the costs and benefits of regulating emissions of carbon dioxide and other greenhouse gases.

DEMONSTRATIONS AND CASE STUDIES

Rising and Falling Water Tables and Climate Change

Because groundwater systems are so sensitive to rates of natural recharge, they are highly responsive to changes in climate. If a prolonged wet period were to occur over several years or even decades, the water table could rise several meters or more as a result of the increased rate of recharge. If a prolonged drought were to occur, the water table could drop several meters or more as a result of the decreased rate of recharge. Most droughts in the past century or so have lasted only several years, and so groundwater levels have not been greatly affected. Recent studies, however, indicate that droughts lasting as long as 220 years have occurred in centuries past, causing groundwater levels to fall far below those of the 20th century. These droughts were associated with global climatic events, such as the Middle Ages Warm Period in the 13th century (see Chapter 13 in the textbook).

In the western United States, two prolonged droughts occurred during the Middle Ages Warm Period, one from 892 to 1112 A.D. and the other from 1209 to 1350 A.D. Both droughts caused springs, lakes, wetlands, and rivers to dry up as water tables fell and natural discharge areas diminished in size and number. If recharge rates to groundwater systems are reduced, the discharge rates will also be reduced. During the prolonged western droughts, forests grew where surface water now exists; for example, in the beds of such rivers as the West Walker, which drains the eastern Sierra Nevada, California (see Figure 12.1 in the textbook). Recent droughts lasting less than a decade exposed the dead stumps of these ancient forests, which were radiometrically dated with carbon. Counting the rings of the trees enabled researchers to determine the durations of the dry periods.

Recharge rates were high enough during glacial periods, such as the most recent one between 35,000 and 10,000 years ago, that many intermountain basins in the American West filled with lake water as groundwater levels rose upward. As areas of natural groundwater increased and lake levels climbed the hillsides, water spilled out and flowed into adjacent basins, forming a chain of lakes that extended from the Canadian border nearly to Mexico. Even in the Grand Desierto of northern Mexico, where modern precipitation is less than 5 cm per year and groundwater levels are very deep, lakes filled the bowls of volcanic craters during this wet period.

Global Warming

See W. H. Freeman's Global Warming web site at

http://www.whfreeman.com/globalwarming

for a discussion of this important environmental issue. The site includes an essay and links to web sites related to global warming.